WATER REUSE: UNDERSTANDING PUBLIC PERCEPTION AND PARTICIPATION

by:

Troy W. Hartley, Ph.D.

RESOLVE, Inc.

Washington, DC

Acknowledgments

National Water Research Institute

American Water Works Association Research Foundation

WateReuse Foundation

2003

Water Environment
Research Foundation

The Water Environment Research Foundation, a not-for-profit organization, funds and manages water quality research for its subscribers through a diverse public-private partnership between municipal utilities, corporations, academia, industry, and the federal government. WERF subscribers include municipal and regional water and wastewater utilities, industrial corporations, environmental engineering firms, and others that share a commitment to cost-effective water quality solutions. WERF is dedicated to advancing science and technology addressing water quality issues as they impact water resources, the atmosphere, the lands, and quality of life.

For more information, contact:
Water Environment Research Foundation
635 Slaters Lane
Alexandria, VA 22314
Tel: (703) 684-2470
Fax: (703) 299-0742
www.werf.org
werf@werf.org

This report was co-published by the following organizations. For non-subscriber sales information, contact:

Water Environment Federation
601 Wythe Street
Alexandria, VA 22314-1994
Tel: (800) 666-0206
Tel: (703) 684-2452
Fax: (703) 684-2492
www.wef.org
pubs@wef.org

IWA Publishing
Alliance House, 12 Caxton Street
London SW1H 0QS, United Kingdom
Tel: +44 (0) 20 7654 5500
Fax: +44 (0) 20 7654 5555
www.iwapublishing.com
publications@iwap.co.uk

Library of Congress Catalog Card Number: 2003100314

WERF ISBN: 978-1-893664-77-7 IWAP ISBN: 978-1-84339-669-7 WEF ISBN: 978-1-57278-178-8

This report was prepared by the organization(s) named below as an account of work sponsored by the Water Environment Research Foundation (WERF). Neither WERF, members of WERF, the organization(s) named below, nor any person acting on their behalf: (a) makes any warranty, express or implied, with respect to the use of any information, apparatus, method, or process disclosed in this report or that such use may not infringe on privately owned rights; or (b) assumes any liabilities with respect to the use of, or for damages resulting from the use of, any information, apparatus, method, or process disclosed in this report.

RESOLVE, Inc., Washington, DC

This document was reviewed by a panel of independent experts selected by WERF. Mention of trade names or commercial products does not constitute WERF endorsement or recommendations for use. Similarly, omission of products or trade names indicates nothing concerning WERF's positions regarding product effectiveness or applicability.

ACKNOWLEDGMENTS

The project team would like to acknowledge and thank many individuals and organizations that contributed to this final product. The three case studies in this report reflect the commitment of time, information, and resources of local and state government agencies, non-profit and citizen organizations, and individuals. The symposium was comprised of 38 participants holding diverse perspectives on nonpotable and potable water reuse and/or the social science underlying our understanding of public perceptions and participation. All participants contributed substantively to the findings reported herein. In addition, staff at all of the project team organizations made critical contributions to the work and are to be thanked for their outstanding support. Finally, this project was funded by the Water Environment Research Foundation (WERF), in cooperation with the National Water Research Institute (NWRI), American Water Works Association Research Foundation (AWWARF), and the WateReuse Foundation.

Report Preparation

Principal Investigator:
Troy W. Hartley, Ph.D., RESOLVE, Inc.

Project Subcommittee

James Crook, Ph.D., P.E. Chair
CH2M Hill

William Bloomquist, Ph.D.
Indiana University/Purdue University - Indianapolis

Paul Gagliardo, P.E.
City of San Diego

Margot Garcia, Ph.D.
Virginia Commonwealth University

Ronald Linsky
National Water Research Institute

David Requa, P.E.
Dublin San Ramon Services District, California

Jeffery Stone
California Department of Public Health

Michael Wehner
Orange County Water District, California

Ronald Wildermuth
Orange County Water District, California

Advisory Group

Maureen Austin, Ph.D., *University of Michigan*
Juliana Birkhoff, Ph.D., *RESOLVE, Inc.*
Caron Chess, Ph.D., *Rutgers University*
James Conklin, *Information Development and Training, Inc.*
Mary F. McDaniel, D.O., J.D., M.P.H., *McDaniel Lambert, Inc.*
Edward Means, *McGuire Environmental Consultants, Inc.*
Michael McGuire, Ph.D., *McGuire Environmental Consultants, Inc.*
Chris J. Wiant, Ph.D., *Caring for Colorado Foundation*

Symposium Participants and Contributors

Robert Adamski, *Gannett Fleming*
Mauren Austin, Ph.D., *University of Michigan*
Bonnie Bailey, *WERF*
Bob Bastian, Ph.D., *U.S. Environmental Protection Agency*
Thomas C. Beierle, Ph.D., *Resources for the Future*
Juliana Birkhoff, Ph.D., *RESOLVE, Inc.*
Huali Chai, *citizen representative, San Jose, California*
Caron Chess, Ph.D., *Rutgers University*
James Conklin, *Information Development and Training, Inc.*
James Crook, Ph.D., P.E., *CH2M Hill*
David Fischer, *former mayor St. Petersburg, Florida*
Paul Gagliardo, P.E., *engineer, City of San Diego*
Margot Garcia, Ph.D., AICP, *Virginia Commonwealth University*
Sam Hadeed, *National Biosolids Partnership, Water Environment Federation*
Troy Hartley, Ph.D., *RESOLVE, Inc.*
Ronald B. Linsky, *National Water Research Institute*
Lorraine Loken, *Water Environment Federation*
Linda Macpherson, *CH2M Hill*
Dale Manty, Ph.D., *U.S. Environmental Protection Agency*
June Marks, *Flinders University, Adelaide, Australia*
Victoria Martz, *Palmer Course Design, Golf Course Architect*
Mary F. McDaniel, D.O., J.D., M.P.H., *McDaniel Lambert, Inc.*
Michael McGuire, Ph.D., *McGuire Environmental Consultants*
Robert Meadow, Ph.D., *Decision Research*
Edward Means, *McGuire Environmental Consultants*
Daniel Okun, Ph.D., *University of North Carolina*
Charlie Rittenhouse, *Citizen Advisory Board, Gwinnett County, Georgia*
John Ruetten, *Resource Trends, Inc*
Ron Seder, *Lake Lanier Association (LLA), Georgia*
Bruce J. Stedman, *RESOLVE, Inc. (facilitator)*
Frank Stephens, *Engineering, Construction, and Process Control, Gwinnett County, GA*
Jeff Stone, *California Department of Public Health*
Lynn Thorp, *Clean Water Action*
Evelyn Torres, Ph.D., *Upper Occoquan Sewage Authority, Centerville, VA*
Mike Wehner, *Orange County Water District, CA*
Chris J. Wiant, Ph.D., *Caring for Colorado Foundation*
Ronald Wildermuth, *Orange County Water District, CA*
David York, Ph.D., P.E., *Florida Department of Environmental Protection*

Water Environment Research Foundation Staff

Research Program Director: Jami Montgomery

Project Manager: Bonnie Bailey

ABSTRACT

Nonpotable and potable (principally in-direct potable) water reuse initiatives in the United States have faced increasing public opposition. Several high-profile initiatives have been halted after several years of planning and tremendous expenditures. To understand why the public holds the perceptions they do and what public participation options exist to address water reuse more constructively, a multidisciplinary analysis was undertaken by a team of social scientists, engineers, and water professionals. Through a comprehensive literature review, three in-depth case studies, and a two-day interactive symposium this framework was developed for water professionals. The framework summarizes five underlying principles that contribute to shaping public perception and acts as a guide for water professionals in their selection of public outreach, education, and participation activities. Adhering to the principles outlined in this report contributes to building public confidence and trust, which in turn helps water utilities engage constructively with the public on challenging, contentious issues. The five principles are:

- Manage information for all
- Maintain individual motivation and demonstrate organizational commitment
- Promote communication and public dialog
- Ensure fair and sound decision making and decisions
- Build and maintain trust

However, no checklist of "to-dos" exists for establishing public confidence and trust. Quite the opposite, this research suggests that a one-size-fits-all model cannot work because the most appropriate ways to achieve the principles can vary from case to case. Thus, the framework includes an analytical structure to assess the community in which a water reuse initiative is underway. Using diagnostic questions and analytical techniques, a comprehensive picture of the community can be generated and monitored over time. Through application of the diagnostic tools and a commitment to the principles outlined above, water professionals can build the public confidence and trust they need to engage with the public on difficult water reuse issues.

BENEFITS

- Improves awareness of the core principles that contribute to shaping public perception and the nature of public participation in nonpotable and potable water reuse initiatives, and in turn, the public's confidence and trust in the wastewater treatment community.
- Advances understanding of the role that public perception and participation can play in building public confidence in the wastewater treatment community, improves the quality of the public dialog on water reuse, and helps determine whether water reuse projects are implemented.
- Provides diagnostic tools and strategies to assess and apply principles to a specific community's water reuse initiatives.
- Supplies tips and strategies for addressing the core principles in water reuse initiatives.
- Provides references and resources for additional information, tools, and techniques.

Keywords: public perception, public participation, water reuse, information, motivation, communication, public relations, fairness, trust, commitment

OPEN LETTER TO WATER SUPPLY AND WASTEWATER UTILITY MANAGERS AND BOARDS OF DIRECTORS

Dear Water Supply and Wastewater Utility Managers and Boards of Directors:

You are part of a remarkable success story. You and your colleagues have dramatically improved water quality and attained a high-level of public trustworthiness, credibility, and excellence. However, in the present social and political climate, public scrutiny about your decision-making processes is higher than ever. Today, there are *real costs* to failing to maintain the public's belief in water resource management and the wastewater treatment community. For example:

- Hundreds of millions of dollars were spent on a 5+ year San Diego water reuse initiative that was halted because of intense public opposition. Other efforts in planned indirect potable reuse in Florida and California have been deferred or rejected by decision makers due in part to active public opposition and the lack of public support.
- Over $500 million was spent on a state-of-the-art wastewater plant that is now undergoing legal challenge by opponents to the project because of its proposed discharge into a reservoir that supplies drinking water to millions of people in Georgia.

These examples supply critical lessons on how to re-establish public confidence. Sustainable water resource management in the 21st century will demand both state-of-the-art technical facilities *and* state-of-the-art public communication and relationships among citizens, interest groups, and the wastewater treatment community.

We are supplying you with this *Water Reuse: Understanding Public Perception and Participation* to stir and guide your thinking about how to further build and maintain public confidence and relationships. The guidance in framework draws upon what many other natural resources, environmental, and engineering agencies have learned over the past three decades. However, because of your history of success and public confidence, the intense public scrutiny and conflict is relatively new in the wastewater treatment context. Thus, the application of this guidance must be viewed as an evolving set of tools based upon empirically identified core principles. The core principles include:

- Manage information for all
- Maintain individual motivation and demonstrate organizational commitment
- Promote communication and public dialog
- Ensure fair and sound decision making and decisions
- Build and maintain trust

The framework truly is guidance, *not* a checklist of "to-dos" to establish public confidence. Quite the opposite, the research suggests that a one-size-fits-all approach cannot work because the most appropriate ways to adhere to the core principles can vary from case to case. Rather, what is critical is the utility's commitment to the principles and its capabilities to tailor the public outreach, education, marketing, and participation effort based upon the context of the community, the capabilities of the participants, and the project. Context and capabilities change over time. Consequently, the tools that are most successful also must change over time.

We believe that you can achieve an unparalleled level of public confidence and be viewed by the public as an essential component of sustainable water resource management in America. However, to reach this level of public confidence, you must be committed to building and maintaining your utility's relationship with the public. While there are no quick fixes, this framework is a resource to help you build public confidence and trust.

Sincerely,

Troy W. Hartley, Ph.D., RESOLVE, Inc.
Principal Investigator

TABLE OF CONTENTS

Acknowledgments ... iii
Abstract, Benefits, and Keywords ... vi
Open Letter to Water Supply and Wastewater Utility Managers and Boards of Directors ... vii
List of Tables ... xi
List of Figures ... xii

Executive Summary ... ES-1
ES1 Project Objectives ... ES-1
ES2 Research Approach ... ES-1
ES3 Research Findings ... ES-2
ES3.1 Principle 1: Manage Information for All ... ES-2
ES3.2 Principle 2: Maintain Individual Motivation and Demonstrate Organizational Commitment ... ES-2
ES3.3 Principle 3: Promote Communication and Public Dialog ... ES-3
ES3.4 Principle 4: Ensure Fair and Sound Decision Making and Decisions ... ES-3
ES3.5 Principle 5: Build and Maintain Trust ... ES-3
ES4 Organization of the Report ... ES-3

1.0 Introduction: People and Water ... 1-1
1.1 Today's Water Reuse Decision-Making Context ... 1-1
1.2 Nonpotable and Potable Reuse ... 1-4
1.3 Framework Development – Research Methods ... 1-4
1.4 Roadmap to the Framework Guidance ... 1-5

2.0 The People Side of Public Decisions ... 2-1
2.1 Societal and Institutional Context ... 2-1
2.2 Group Profile and Decision Context ... 2-3
2.3 Individual Characteristics ... 2-3
2.4 Conclusion ... 2-5

3.0 Building and Maintaining Public Confidence: The Core Principles ... 3-1
3.1 What is Success in Public Participation and in Perception of Water Reuse? ... 3-2
3.2 Principle 1: Manage Information For All ... 3-3
3.3 Principle 2: Maintain Individual Motivation and Demonstrate Organizational Commitment ... 3-6
3.4 Principle 3: Promote Communication and Public Dialog ... 3-9
3.5 Principle 4: Ensure Fair and Sound Decision Making and Decisions ... 3-11
3.6 Principle 5: Build and Maintain Trust ... 3-12
3.7 Conclusion ... 3-14

4.0 Putting the Principles into Action: Diagnostic Tools to Help Tailor Public Outreach, Education, Marketing, and Participation Products and Activities ... 4-1
4.1 What Are Diagnostic Tools? ... 4-1
4.2 Societal and Institutional Context ... 4-2
4.3 Group Profile and Decision Context ... 4-4

4.4 Individual Characteristics 4-6
4.5 What Do I Do Now? Analyzing Community Assessment Data 4-7

5.0 Putting the Principles into Action: Tips from "People" Specialists 5-1
5.1 Consensus-Building Processes 5-2
5.2 Leadership and Politics 5-4
5.3 Change Management and How People Deal With Behavioral Change 5-4
5.4 Sharing Power While Maintaining Decision-Making Authority 5-5
5.5 Psychology and Information Sharing 5-6
5.6 Risk Communication and Motivating People to Engage 5-7
5.7 Conclusion 5-8

6.0 Putting the Principles into Action: Public Participation Techniques 6-1
6.1 Approaches to Public Participation 6-1
6.2 Marketing Reclaimed Water 6-14
6.3 Advisory Committees – Tips from a Former City Council Member 6-14
6.3.1 Representation 6-15
6.3.2 Leadership 6-15
6.3.3 Length of Term 6-15
6.3.4 Agenda Setting 6-16
6.3.5 Final Report 6-16
6.4 Conclusion 6-17

7.0 Conclusion 7-1
7.1 Principle 1: Manage Information For All 7-2
7.2 Principle 2: Maintain Individual Motivation and Demonstrate Organizational Commitment 7-2
7.3 Principle 3: Promote Communication and Public Dialog 7-2
7.4 Principle 4: Ensure Fair and Sound Decision Making and Decisions 7-3
7.5 Principle 5: Build and Maintain Trust 7-3
7.6 Additional Resources 7-3
7.6.1 Public Participation and Other Guidance Documents 7-3
7.6.2 Web Sites 7-4
7.7 The Next Steps 7-5

Appendices
A. Public Perception and Participation in Water Reuse: Literature Summary A-1
B. Public Perception and Participation in Water Reuse: Case Summaries B-1
C. Public Perception and Participation in Water Reuse: White Papers C-1

References R-1

LIST OF TABLES

1-1 Challenges and Opportunities Faced by Water Reuse Professionals 1-4
4-1 Diagnostic Questions: Societal and Institutional Context 4-3
4-2 Diagnostic Questions: Group Profile and Decision Context 4-5
4-3 Diagnostic Questions: Individual Context 4-6
5-1 Addressing Complex Issues in Consensus-Building Processes 5-3
5-2 Tips for Effective Communication 5-7
6-1 The IAP2 Public Participation Spectrum 6-3
6-2 IAP2 Public Participation Toolbox: Passive Public Information Techniques 6-4
6-3 IAP2 Public Participation Toolbox: Active Public Information Techniques 6-6
6-4 IAP2 Public Participation Toolbox: Small Group Public Input Techniques 6-9
6-5 IAP2 Public Participation Toolbox: Large Group Public Input Techniques 6-10
6-6 IAP2 Public Participation Toolbox: Small Group Problem-Solving Techniques 6-11
6-7 IAP2 Public Participation Toolbox: Large Group Problem-Solving Techniques 6-13

LIST OF FIGURES

3-1 Elements of Success in Public Decision-Making and Participation 3-2
3-2 The Complete Water Cycle 3-5
3-3 Growing a Solid Tree of Public Confidence 3-14

EXECUTIVE SUMMARY

ES.1 Project Objectives

1. Provides guidance for water resource professionals to incorporate stakeholder priorities in water reclamation programs.
2. Helps members of the water industry address the social and political complexity of adopting potable and water reuse and recycling as part of a sustainable community strategy.

ES.2 Research Approach

This framework was developed through a collaborative, multidisciplinary approach that sought to integrate diverse social science perspectives and practical utility management experiences. A multidisciplinary investigative team, comprised of social scientists from psychology, sociology, public policy, planning, communication, and organizational studies, along with engineers and water utility managers, provided input in a three-step process.

First, an extensive background research effort was conducted, consisting of a literature review, three in-depth case studies, and 10 reflective concept papers from each of the scientific and professional perspectives.

Second, the background materials were used to prepare a selective group of participants for a facilitated, interactive symposium, held in August 2001 in Washington, DC. Participants included utility managers, engineers, citizen leaders, elected officials, social scientists, and representatives from the regulatory, public relations, and risk communication sectors. The participants carefully examined and thoroughly discussed the findings from the literature and case studies. They explored specific strategies for addressing public perception and participation issues. The symposium participants also considered the best application and design of the final framework report.

Third, the framework report was developed by the investigation team with extensive peer review. The framework underwent external review by the symposium participants; an advisory group comprised of social scientists who specialize in public perception and participation issues and environmental engineers with experience in public participation; and the Water Environment Research Foundation Project Subcommittee, comprised of social scientists, utility officials, and state regulatory officials.

ES.3 Research Findings

The framework presents a strategy for building public confidence and trust through a principle-based approach to public outreach, education, and participation. It requires two steps:

1. The application of diagnostic tools to generate a comprehensive community assessment and the implementation of an ongoing community assessment monitoring program
2. The tailoring of specific public outreach, education and participation activities based upon the outcome of the community assessments *and* consistent with the five underlying principles (see below)

The diagnostic step calls for an analysis of contextual factors specific to your community, along with a scale from societal to individual levels. Specifically, the contextual factors include:

- Societal and institutional context
- Group profile and decision context
- Individuals' characteristics

Analysis of data collected on these contextual factors aims to identify important patterns and trends. It is the analysis of the contextual data that enables the tailoring of products and activities.

Second, once an understanding of the context exists, and a monitoring system is in place to help detect changes in the context, the five principles are applied to the context to determine specific public outreach, education, marketing and participation strategies, tools, and techniques. This exercise can be applied at several levels of management from selecting overall communication strategies to designing and implementing individual participation activities (e.g., a citizens' advisory committees or a marketing message). The framework principles are:

ES.3.1 Principle 1: Manage Information for All

Diverse information should be managed to promote learning, communication, and mutual understanding among all the stakeholders, given that different people learn and communicate differently.

- Ensure equal access to information.
- Provide multiple methods of information presentation and communication.
- Support and adhere to individuals' information processes and tendencies.

ES.3.2 Principle 2: Maintain Individual Motivation and Demonstrate Organizational Commitment

Individuals should find more good reasons (e.g., benefits for self and community) than bad reasons (e.g., real or perceived risks) to engage in public dialog. Organizations should demonstrate genuine commitment to engaging the public.

- Ensure and nurture multiple motives for the public to engage, including individual and community benefits.
- Demonstrate genuine organizational commitment to hearing the public and its concerns.

ES.3.3 Principle 3: Promote Communication and Public Dialog

Communication and the broader, cumulative public dialog should take multiple forms or venues and be ongoing at all stages of a decision-making process.

- Expect to have to repeat the message multiple times (and in multiple ways) before it is understood by others.
- Expect to have to listen to others' messages multiple times and in multiple ways before you understanding them.
- The quality of the public dialog is an indicator of the public confidence and relationship.

ES.3.4 Principle 4: Ensure Fair and Sound Decision Making and Decisions

Both the decision-making process and the outcome should be perceived to be fair and sound by participants.

- Fairness should include process or procedural dimensions and outcome or distributive elements.
- Sound decisions are reasoned, well thought out, and based upon accepted knowledge

ES.3.5 Principle 5: Build and Maintain Trust

The cumulative attention to the previous principles contributes to the establishment and maintenance of public confidence and trust.

- Build trust and credibility reserves with the public and tap on those reserves when necessary.

ES.4 Organization of Report

- *Chapter 1.0, Introduction: People and Water* introduces the topic of water reuse and public perception, chronicling the literature within the water reuse field and describing the approach taken by the research team.
- *Chapter 2.0, The People Side of Public Decisions* discusses challenging issues and complex questions faced by water professionals working within the social context of public decision making.
- *Chapter 3.0, Building and Maintaining Public Confidence: The Core Principles of the Framework* explores each of the core principles in greater detail and provides examples of how the principles have played out in particular cases and professional experiences. While the core principles are vital to understanding what shapes public perception and the nature of participation, it is the principles in action through particular activities and tasks that are of great interest to many water professionals.
- *Chapter 4.0, Putting the Principles into Action: Diagnostic Tools to Help Tailor Public Outreach, Education, Marketing, and Participation Products and Activities* provides hints, tools, and guidance for water professionals.
- *Chapter 5.0, Putting the Principles into Action: Tips from "People" Specialists and Common Public Participation Strategies* details suggestions from experienced professionals and researchers.

- *Chapter 6.0, Putting the Principles into Action: Public Participation Techniques* describes successful public participation approaches. These strategies can be starting points for tailoring approaches to a specific community.
- *Chapter 7.0, Conclusion* pulls the themes together, emphasizing both existing opportunities for the wastewater treatment community to achieve a high level of public confidence and the critical next steps that the social science community must undertake to answer many remaining questions about public perception and participation. Chapter 7.0 also includes resources for additional information.

CHAPTER 1.0

INTRODUCTION: PEOPLE AND WATER

Finding solutions to the assortment of technical and scientific issues and problems associated with designing and building wastewater treatment and water reuse systems can be challenging and even fun for environmental engineers and utility managers. However, rarely do you hear utility professionals say that solving "people" issues is fun. Nonetheless, the people side of water reuse decisions can be equally and often more challenging than solving the technical or scientific issues. People are complex and often come across as perplexing to public decision makers. As the Water Environment Federation report *Using Reclaimed Water to Augment Potable Water Resources* (1998, p. 231) noted:

> The ability to gauge the full scope of underlying issues is also critical to a project's final acceptance by the public. Although technological considerations may open the door to dialog, leading a community through the door requires mobilizing interest and thoughtfully managing the process. Bare facts and technical assurances may not guarantee the success of a … reuse project. Although there are tricks of the trade such as avoidance of jargon and preparedness, a process may also require basic character strengths of patience, empathy, respect, wisdom, sincerity, and humor. The ability to apply these characteristics comes from stamina and experience.

Over the last 20 to 30 years, social scientists and public outreach, education, and participation professionals have been learning a lot about what makes people tick in public decision-making processes. In addition, they understand what contributes to making public discussions challenging. This chapter introduces the landscape of water reuse and the approach taken to develop this framework.

1.1 Today's Water Reuse Decision-Making Context

Water can be a limited resource in a rapidly expanding global population. Many water resource professionals believe that reclaiming water after it is treated in a modern wastewater treatment plant is an important and underutilized element of sustainable water resource management. Water reuse for nonpotable (e.g., irrigation, industrial) or indirect potable (e.g., discharge into drinking water reservoirs) purposes has been considered across the country, but particularly in drier or drought-ridden areas, such as Arizona, California, Colorado, and Texas, or areas that are experiencing rapid population and economic growth that strain water supplies (e.g., Georgia and Florida). It is only a matter of time before many other communities consider water reuse options.

Survey and case study research since the 1970s has found that much of the public supports the concept of using reclaimed water and has been somewhat supportive of nonpotable reuse initiatives (U.S. Environmental Protection Agency [EPA], 1992). People generally favor reuse that promotes water conservation, provides environmental protection benefits, protects human health, and cost effectively treats and distributes a valuable and limited resource. However, as the water options become more concrete (i.e., with specific proposed projects in their communities and the likelihood of human contact with reused water increasing) attitudes change, and public support wanes (Bruvold, 1988).

Over the last couple of years, consumer surveys have been conducted in Orange County, California, where a water reuse initiative is underway. The public was concerned because reclaimed water was originally wastewater (Lawrence, 2000). A survey in 2000 in the United Kingdom corroborated the existence of a "source" factor. It found people are more willing to use recycled water from their own wastewater than from second parties or a common public source (Jeffery, 2001). In other words, the "yuck factor," the term that water professionals use to discuss the cumulative impact on public perception of water reuse, may be tempered by an individual's proximity to the waste source. The factors contribute to the degree of public acceptance of water reuse, according to survey and case study research from the late 1970s through early 2000 are summarized below. Public acceptance seems to be higher when:

- The degree of human contact is minimal.
- Protection of public health is clear.
- Protection of the environment is a benefit of reuse.
- Promotion of water conservation is a benefit of reuse.
- The cost of treatment and distribution technologies and systems is reasonable.
- Perception of wastewater as the source of reclaimed water is minimal.
- The community has high awareness of water supply problems.
- The role of reclaimed water in the overall water supply is clear.
- The perception of the quality of the reclaimed water is high.
- Confidence in local management of public utilities and technologies is high (Bruvold, 1988; Lawrence, 2000; Lohman, 1984; Jeffery, 2001; and U.S. EPA, 1992).

The social, political, and economic context in which water reuse decisions are occurring is muddied by immense challenges and unique opportunities. In addition to site-specific factors that can shape the public's perception and the nature of their participation in water reuse decision making, some significant national and regional trends also exist.

In general, trust and confidence in public agencies and officials is in decline (Putnam, 1995, The Pew Research Center for People and the Press, 2001). This appears also to be true for the wastewater utilities. Likewise, the belief that even the best technologies can remove all impurities and germs from wastewater is in decline, at least in California (Lawrence, 2000).

Furthermore, while the public has reported trusting university-based scientists and the medical community on technical and health issues related to water reuse, preliminary survey evidence shows that people trust their impressions of water quality (often based upon the water's cloudiness or turbidity) more than they trust the experts (Broad, 1996; Jeffery, 2001).

In the 1990s, a number of high-profile indirect potable water reuse projects encountered stiff public opposition. The public effectively organized into influential political interest groups and prevented the implementation of water reuse projects. Common terms in the public dialog included "toilet to tap" and "sewage beverage." In San Diego, the newspaper published a cartoon of a dog drinking from a toilet and a man behind the dog saying, "Move over." As one veteran of risk communication in California water resource decisions said, "discourse is often dominated by a few true believers on both sides, often with extreme views, and is usually competitive and antagonistic" (McDaniel, 2001). Furthermore, members of the scientific and technical community disagree over the public-health viability of indirect potable water reuse, with major water resource professional associations and respected research and expert panels taking opposing views (National Research Council, 1998; Blue Ribbon Panel, 1998).

In one California market study, information sharing, educational activities, and providing time to reflect upon the concept of water reuse before stating an opinion were shown to increase support. However, they also intensified the extremes. Those who are opposed to water reuse became more strongly opposed, and those who support it became more strongly supportive (Lawrence, 2000; Broad, 1996).

Nevertheless, there is good news for the water professional. In spite of serious opposition in some California cases, the public has expressed an interest in meaningful involvement in water reuse decision making and finding solutions to ensure an independent and secure water supply for their communities (Bruvold, 1981 and 1991; Lawrence, 2000). The public is, in a general sense, aware of water supply problems in many parts of the country, and a few believe that some form of potable reuse is inevitable, given growth and water supply constraints (Lawrence, 2000; Broad, 1996). Table 1-1 summarizes the challenges and opportunities water reuse professionals face.

Table 1-1. Challenges and Opportunities Faced by Water Reuse Professionals

CHALLENGES	OPPORTUNITIES
♦ Public trust and confidence in public agencies and officials has declined.	♦ The public is interest in being meaningfully involved in water reuse decisions.
♦ The belief that best technologies can remove all impurities and germs from wastewater has declined.	♦ The public is interested in finding ways to ensure independent and secure water supplies for their communities.
♦ While the public tends to trust university-based scientists and the medical community on technical and health issues, they trust their impressions of water quality more.	♦ While the public is not well versed in the water cycle, they generally are aware that water supply problems exist in many parts of the country.
♦ The public impression of water quality often is based upon the water's turbidity.	♦ The public believes that some form of potable reuse is inevitable, given growth and water supply constraints.
♦ While education and outreach activities can increase support, they can also intensify the extremes. Those who oppose become more strongly opposed, and those who support become more strongly supportive.	♦ Information sharing, educational activities, and opportunities for reflection upon the concepts of water reuse can increase support.
Source: Broad, 1996; Bruvold, 1981 and 1991; Lawrence, 2000; Jeffery, 2001; Putnam, 1995; The Pew Research Center for People and the Press, 2001	

1.2 Nonpotable and Potable Reuse

The intensity of the public's reaction, concern, and debate over water reuse is magnified when the reuse issues change from nonpotable to potable. In fact, while many in the scientific and engineering communities believe that nonpotable reuse is feasible and often desirable, the acceptance within the scientific and technical communities is far from uniform in regards to indirect potable reuse. The technical and scientific challenges and difference of opinion among scientists and engineers introduces greater uncertainty into the public debate. The uncertainty can be accompanied by more intensity in the opposition and expression of public concerns. While the principles discussed here are the same for a public discussion about nonpotable or indirect potable water reuses, underlying public interests, values, and concerns may be more strongly expressed in an indirect water reuse case. As a result, the public discussion may be more contentious, and the application of the principles even more important in promoting a constructive public dialog.

1.3 Framework Development – Research Methods

Within this challenging and dynamic social and political context, the water reuse community has asked itself why the public holds the perceptions of reuse they do and what can be done to discuss and address these perceptions through public participation efforts. The Water

Environment Research Foundation (WERF), in cooperation with the National Water Research Institute, American Water Works Association Research Foundation, and the Water Reuse Foundation, funded this project as the first step into the social science research on perception and participation and its application in a water reuse context.

This framework was developed through a collaborative, multidisciplinary approach that sought to integrate diverse social science perspectives *and* practical utility management experience. A multidisciplinary investigative team, comprised of social scientists from psychology, sociology, public policy, planning, communication, and organizational studies, along with engineers and water utility managers, provided input in a three-step process.

First, an extensive background research effort was conducted, consisting of a literature review, three in-depth case studies, and 10 reflective concept papers from each of the scientific and professional perspectives.

Second, the background materials were used to prepare a selective group of participants in a facilitated, interactive symposium, held in August 2001 in Washington, DC. Participants included utility managers, engineers, citizen leaders, elected officials, social scientists, and representatives from the regulatory, public relations, and risk communication sectors (see list of participants in *Acknowledgments* section). The participants carefully examined and thoroughly discussed the findings from the literature and case studies. They explored specific strategies for addressing public perception and participation issues. The symposium participants also considered the best application and design of a final framework report.

Third, the framework report was developed by the investigative team. This included extensive peer review. The framework underwent external review by the symposium participants; an advisory group comprised of social scientists who specialize in public perception and participation issues and environmental engineers who have experience in public participation; and the WERF project subcommittee, comprised of social scientists, utility officials, and state regulatory officials.

1.4 Road Map to the Framework Guidance

The framework is a guidance document for water professionals who are building and maintaining public confidence and relationships. The principles described in the framework build upon what many other public natural resources, environmental, and engineering agencies have learned over the past three decades. However, the framework's application in the wastewater treatment and water reuse context is relatively new because historically public confidence in water resource management has been high and only recently has there been heightened public scrutiny and conflict. Consequently, the framework must be viewed as an evolving tools based upon empirically identified core principles. The core principles include:

- Manage information for all
- Maintain individual motivation and demonstrate organizational commitment
- Promote communication and public dialog
- Ensure fair and sound decision making and decisions
- Build and maintain trust

CHAPTER 2.0

THE PEOPLE SIDE OF PUBLIC DECISIONS

There are many societal factors that shape the public context in which water reuse decisions take place. These factors exist on the societal, organizational, process, group, and individual levels, ranging from inherent societal conflict, through statutory authority and group norms, to an individual's cognitive functioning. While the broad societal factors contribute to a context that can feel frustrating and cumbersome to water professionals, America's democratic society is structured to make decisions through a managed and constructive debate (although not always efficient) amongst multiple perspectives and values.

Conflict is natural. It should not be a surprise when it surfaces in water reuse initiatives, nor should it be a surprise that the conflict raises social issues from long-seeded cultural, ethnic, or socioeconomic disputes that involve power imbalances. Likewise, powerful legal and institutional structures define the nature of public participation programs and often the scope of the decision-making authority. The institutional decision-making processes and the group dynamics that emerge from the interaction of people often introduce further challenges to participation. Also powerful and resistant to change, organizational cultures and professional norms can guide behavior and interactions between individuals in government, business, non-profit organizations, and the public. Finally, deeply embedded cognitive practices can influence how people manage information, engage in public-participation processes, and function in public discussion.

While the previous chapter included a summary of some of the people factors in water reuse, this chapter summarizes the broader social challenges and complex questions faced by water professionals. These issues are not specific to water reuse, but are faced by all public decision makers.

It is within this broader social context of public decision making that water professionals are building or damaging public confidence and trust and defining the overall relationship between the utility, the citizens, and the community leaders. This chapter lays the foundation of what shapes the social context in which decisions take place.

2.1 Societal and Institutional Context

America is a diverse society with many socioeconomic, ethnic, and cultural differences that include a range of values and beliefs. Within this demographic diversity are differences in power and authority, as individuals, groups, and communities have varying financial and human resources, organizational and technical capacity, access to information, and social networks. Furthermore, often long, historical relationships exist between groups that have varying degrees of power. The consequences of a community's diversity and the historical relationships can be entrenched in the societal norms, networks, and social trust that comprise social capital. The social capital often embellishes the significance of these differences, hardens resistance, and fosters conflict, although occasionally it may promote the acceptance of differences and the empowerment of groups.

More often than not, inherent inequities in current and historical power relationships exist within society, and conflict among these groups is commonplace. While conflict can be challenging and may be quite antagonistic, social conflicts carry out critical and useful functions. They define and express important debates about justice, aesthetics, public health, politics, and economics. The presence of conflict is a sign of a healthy society that is capable of communicating different values.

During environmental decision making, individuals and groups can perceive real threats to their public health, disruptions in their social networks, and significant loss of personal control over the decisions influencing their lives. People can feel disempowered. Responses to disempowerment can often track the historical conflict behaviors and patterns predetermined by the societal power structure – that is, the conflict may likely follow previous divisions among community groups. So it is very likely that water reuse discussions could bring forth many of the underlying concerns associated with societal and community power imbalances and the historical behaviors and patterns of conflict. To the water professional, the debate may appear to be about unrelated social issues and not the environmental decision at hand, but conflict behavior and patterns are normal.

However, while conflict can be inherent and even healthy in society, and environmental decision-making can stir underlying societal power differences, this does not mean that conflict always results in escalation of contentiousness, stalemates among parties that halt decisions and implementation, or even violence. Without constructive interactions and active intervention, conflicts in communities can escalate through predictable stages to hostility between parties. Conflict analysis and resolution theory has identified strategies to productively manage community conflicts (Birkhoff, 2001).

While it requires little elaboration, institutional and decision-making authority in this democracy often is delegated to government agencies by elected legislatures at the local, state, and federal level. In turn, administrative procedures, rules and other process-oriented laws (e.g., National Environmental Policy Act) govern how many elements of public participation programs are implemented. In addition to predefined decision-making authority and scope and mandated decision-making procedures, elected legislatures determine appropriation levels.

These institutional structures define the constraints and opportunities for public participation and the level and type of engagement that can most readily occur between public officials and citizens. In turn, the institutional structures provide many of the tools available to public officials as they seek to build public confidence (Wilson, 1989).

2.2 Group Profile and Decision Context

As people come together to make decisions there are several group dynamics and processes challenges that can impact participation and perception. For example, multiple parties often have diffuse interests, values, and concerns. (Parties can be individual citizens as well as organizations of different sizes with varied memberships and different levels of resources, power, negotiation skills, outreach capacity, experience, and political skills.) This context can magnify the impact of issues that have a high level of technical complexity and scientific uncertainty as well as issues that receive considerable media attention. Under these circumstances, the individual's sense of empowerment and fairness can be undermined.

Organizations and disciplinary professions have their own culture (sometimes more than one) or persistent, patterned way of thinking about tasks and relationships to other people within and outside of their organization and/or profession. Cultures involve strongly held values that are resistant to change. Tasks outside the norms of an organizational or professional culture often are difficult to accomplish effectively or efficiently. Similarly, new relationships can be difficult to establish.

The norms and expectations determined by organizational and professional cultures contribute to how professional staffs perceive themselves, as well as the image and impression they communicate to the public. For example, norms contribute to the specific assumptions and conclusions that experts reach relative to laypersons about what information is important to solving problems, what values are relevant to a decision, and which options are appropriate. Organizational and professional cultures may shape the nature and expectations of communication styles and the distinct "languages" spoken by community members, elected officials, government bureaucrats, attorneys, scientific and technical experts, and other distinct organizational or professional groups (Wilson, 1989).

While it is common to consider organizational and professional norms and subject matter expertise to be a function of governmental agencies and the employees within, norms and attitudes of expertise in the form of local knowledge can also be exhibited by members of community organizations. In other words, community groups or citizens can exhibit the same behaviors and communicative styles that arise from norms and a sense of expertise. Based upon their knowledge, they too have biases about what information is important to solving problems, what values are relevant to a decision, and which options are appropriate.

2.3 Individual Characteristics

A range of individual characteristics shape the public context and landscape in which all public decisions take place. For example, attitudes vary widely and contribute to how individuals perceive information, the issues and solutions they believe to be relevant, and their roles and responsibilities in decision making. Risk perceptions are a type of attitude, and experts seem to view risks differently than lay people, regardless of educational level. The public may be more concerned with the impacts and severity of an accident than the likelihood of that accident occurring, whether the risk is voluntarily or involuntarily imposed upon them. The scientific or technical expert may be less concerned with the severity of risk and the difference of opinions among scientists and more focused on the risk probabilities and relative risks.

Individual ability to articulate, listen, clarify, reflect upon and respond to different information varies greatly. Some people are constructive communicators who can listen well, seek clarification, provide feedback, and model communicative behaviors for others. At the same time, people assess other people as speakers based upon the speakers' sincerity and honesty, legitimacy and credibility, the thoroughness of information, and the accuracy of information.

Some individuals exhibit leadership capabilities. A leader may function as collaborative leader, leveraging the skills of individual members of a group, facilitating group processes, and articulating agreements. Other types of leaders may be able to ensure the quality of communication, clarify the scope of tasks, provide resources, and demonstrate a commitment to the process. Leadership contributes to how issues of power and personal behavior play out in a public dialog.

While there will be a range of interests, concerns, and values among members of a community, there also can be various incentives and motivations for engaging in a public dialog about an issue. Interests and concerns are the individuals' articulated reasons for believing or acting. They may include a set of desires, needs, and expectations that exist in a particular decision context. Values are deeper core principles, desires, and expectations that are often reflected in the interests and concerns that people express. Motivation refers to the reasons why people decide to engage and remain engaged in an issue. People may need more than one motive or incentive to act, including self- and common-interests. Common reasons for acting include:

- Personal enjoyment, satisfaction, and fun from engagement
- Perception of fairness
- Suspicion of others' interests as opposing one's own interests
- Sense of importance of tasks and impact on community
- Hope for the future and optimism that a good solution can be found
- Framing problems as opportunities

Cognitive psychology deals with how people manage information and knowledge in complex networks within the brain. Some psychologists argue that the way in which the brain processes information has been evolutionarily determined. Thus, powerful biases have been hard-wired into the human subconscious from hundreds of thousands of years of experience and evolution. These cognitive functions influence how people perceive information and behave in the face of new and challenging information, as is the case when people react to water reuse initiatives (Kaplan and Kaplan, 1982; Kaplan, Kaplan, and Ryan, 1998).

Repeated and frequent experiences can shape how people think. For example, turning on the water faucet every day and finding plenty of water reinforces the perception that there is plenty to go around, countering messages of water shortages or the need for conservation. In New Jersey, homeowners whose water was supplied by wells were more aware of water supply and management issues that public water supply users because of personal experience. Those who had experienced incidents of contaminated well water were far more aware of water-quality issues than public water users (Chess, 2001).

Another cognitive function that guides human thought deals with images and experiences that are particularly powerful, such as surprises, traumatizing events, scary events, or exhilarating activities. Experiencing powerful events or encountering unusual images can have profound effects upon human perception. Such moments are accompanied by increased awareness. They affect emotions and impact how people think and behave from that point forward. For example, emotional "toilet-to-tap" cartoons and other negative images used in opposition to water reuse initiatives can enhance negative emotions and affect public perception.

In addition, cognitive mechanisms have evolved to promote rapid interpretation of data. For example, when a person is presented with new information or an unfamiliar experience, the normal response is to fit that information into the existing knowledge structure. Often this works well, and information is processed efficiently without any conscious effort. However, it also can produce a tendency to jump to conclusions.

Conversely, taking in too much new or unfamiliar information, particularly if it is presented in an overwhelming or highly technical manner, can lead to confusion and even antagonism. People respond negatively to information overload. In fact, the connection between factual information and emotional affect is hard-wired into the cognitive process. In other words, it is not unusual for people to react emotionally to informational sessions on water reuse, even when the information presented appears to be neutral data. The outcome can be antagonism from citizens who are trying to understand and frustration from the professional staff whose efforts to educate don't appear to be appreciated or to make a difference (Austin, 2001).

2.4 Conclusion

The rich societal, institutional, organizational, professional, and individual setting in which public decisions are made presents an enormously complex landscape on multiple levels. On one hand it underscores the impossibility of effectively managing the public's perceptions or attitudes in a meaningful, long-lasting way. Yet on the other hand, the complex landscape provides multiple venues to address public perception of water reuse. There are many opportunities for water utility professionals to approach the people-side of water reuse. For more details on these, please see the literature review in Appendix A.

CHAPTER 3.0

BUILDING AND MAINTAINING PUBLIC CONFIDENCE: THE CORE PRINCIPLES

As discussed in the previous chapter, public decisions are made within a broad societal structure. It is bounded in part by the societal and institutional context that shape power relationships, by group norms and processes, and by human characteristics. These are powerful forces that shape public participation and the behavior and expectations of all people who participate in public decision making. At the same time, more than 25 years of social science research on public policy making and public participation processes, including extensive work in many fields of natural resource and environmental management, have begun to identify additional contributing factors to public confidence. These include some lessons on what can be done to build and maintain public confidence that are within the control of water professionals.

That is not to say that social scientists have figured it all out. Quite the opposite, significant limitations exist. The literature on public participation and perception is dominated by in-depth case studies and geographically narrow survey research. Only recently have attempts been made to conduct wider analyses across larger samples of case studies or surveys across broad regions within a specific type of environmental topic (e.g., watersheds). A National Research Council effort is underway to uncover results from across the existing empirical research and to structure analytical protocols to promote greater cross-case analysis and general findings.

Nonetheless, the research on water reuse reported here does confirm many conclusions from the public participation and perception literature. These research findings were further validated by similar outcomes that emerged from independently conducted biosolids, odor, and environmental management system studies funded by WERF (WEFTEC, 2001). The findings suggest that there are several core principles that contribute to how public confidence and perception are built and maintained (or damaged and eroded) within a public decision-making process.

The principles are not a checklist of "to-dos" that establish public confidence. Quite the opposite, research suggests that a one-size-fits-all model cannot work because the most appropriate and effective tools and techniques that reflect the core principles vary from case to case. What is critical is a utility's commitment to the tailored application of the principles based upon the context of the community, the capabilities of the participants, and the decision at hand.

The core principles include:

- ♦ Manage information for all
- ♦ Maintain individual motivation and demonstrate organizational commitment
- ♦ Promote communication and public dialog
- ♦ Ensure fair and sound decision making and decisions
- ♦ Build and maintain trust

This chapter discusses each of these core principles in greater detail. Before examining each principle, it is important to consider the issue of success, what defines success in public participation and perception within the water reuse context, and how success is defined in the framework.

3.1 What is Success in Public Participation and in Perception of Water Reuse?

Opinions vary as to what constitutes success in a public participation effort. For some, the outcome of a decision-making process is the key. That is, success is the adoption of a water management plan that includes reuse, implementation of a water reuse project, or the acceptance by the public of a nonpotable or indirect potable reuse project. Many water reuse professionals function under this assumption and define success as implementing a project on time and within budget.

For others, successful public participation is determined by the decision-making process. Under this standard, success relates to fairness, level of representation, information availability, data quality, and other characteristics of a democratic process. To others, success is viewed by what is accomplished beyond the decision at hand. For example, has the process and decision built social capital, mutual respect, and strengthened relationships among participants so that future decisions will be made more easily? Others aim to achieve some level of success on process, outcome, and relationship dimensions simultaneously. The multiple, even conflicting, definitions of successful public participation are not unique to water reuse, as many fields of natural resources and environmental decision making have struggled to find an acceptable definition of success (Chess, 2001; Stern, 2001).

The framework takes the middle ground. It accepts multiple definitions of success and does not provide one correct or appropriate definition. The outcome-oriented definition of success typically used in the water reuse community may be too narrow and may contribute to some conflict, since other participating parties may not share the same outcome criteria and may have process or relationship elements to their own view of success. In reaching any decision, a trade-off among process, outcome, and relationship-building dimensions of success will occur (Figure 3-1).

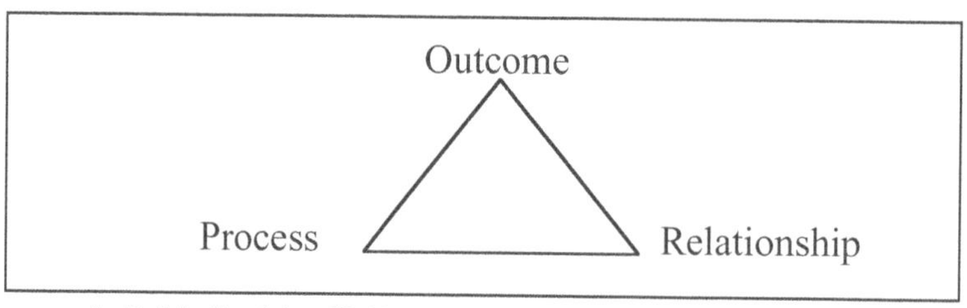

Figure 3-1. Elements of Success in Public Decision Making and Participation.

The framework recommends that all participants define success and that participants share their perspectives with other participants. While conflicting views will emerge and it will not be possible to accommodate all perspectives in a decision-making process or the outcome, honesty about the differences contributes constructively to building public confidence and attends to the core principles advanced by the framework.

3.2 Principle 1: Manage Information for All

Diverse information should be managed to promote learning, communication, and mutual understanding among all the stakeholders, given that different people learn and communicate differently.

- Ensure equal access to information.
- Provide multiple methods of information presentation and communication.
- Support and adhere to individuals' information processes and tendencies.

Managing information for all means promoting communication, providing opportunities to exchange information and develop mutual understanding, and allowing free access information. The complexity of the information should not get in the way of promoting a broad understanding of a variety of information. "Information" in a water reuse decision is defined very broadly in this principle, referring to the science and technologies, local knowledge and site-specific characteristics, values and interests, and the local context (e.g., political, social, economic, and environmental landscape). In other words, information includes the diverse types of data, evidence, values and context that relate to the individuals involved, specific problem and decision-making at hand, and the broader societal realm in which decisions are made and implemented.

The information itself should be managed in a manner that promotes the building of mutual understanding and public confidence. If not done well, this can undermine and damage public confidence and increase misunderstanding and misperceptions. Furthermore, the uncertainty or incompleteness of information also influences perception about water reuse. Therefore, information as discussed here contains concrete facts, as well as flavor or richness that embellishes the influence of those facts.

Although people cognitively function in similar ways, what they know is based upon their unique experience and knowledge. Thus, people communicate and learn differently. While people have considerable shared knowledge, they also possess various information and perspectives. For example, a New York City artist, Wall Street executive, and city wastewater engineer may share a knowledge and passion for Broadway musicals, the Yankees, and the same political party, yet have vastly different knowledge structures in their professions, personal lives, and risk perception.

Equal access to information implies not hiding information from the public. As one risk communication consultant said during the August 2001 symposium on this topic, "People will forgive you if you make a mistake, but they will not forgive you if you do not tell them you made a mistake." Equal access enhances the perception of fairness in the decision-making process and the ultimate decision.

EXAMPLE: In the Gwinnett County, Georgia, case study county officials said that they tried to tell everyone when mistakes were made. "We put the bad news out in front of people… and it buys press credibility and with our oversight board" (Hartley, 2001). A journalist confirmed, "[the government official] is adept about describing technology and he's a straight forward guy, who will be straight with you about any potential problems" (Hartley, 2001).

To account for diverse learning styles, managing information for all also means using *multiple methods* to present information (e.g., written materials, public presentations, public service announcements, seminars and workshops). One communicative approach alone is insufficient because it cannot accommodate all learning styles.

EXAMPLE: Orange County, California, is undertaking an extensive public outreach and education program as part of their groundwater replenishment system initiative. In addition to community research and program evaluation activities, the public education effort includes multiple outreach activities, including: community presentations; cable/local television appearances; project videos and newsletters to libraries, water district customer counters, cable television stations; mailings; media relations program releases, visits, and site tours; bump-in-the-road crisis teams; and public tours of facilities (Wildermuth, 2001).

Furthermore, information sharing techniques should support and be compatible with human *cognitive processes*. That is:

- Present information in small installments. Do not overwhelm the receiver.
- Use common, shared vocabulary rather than jargon.
- Present information in settings familiar to the audience. Go where the people are. Familiar settings can alleviate concerns of looking foolish when learning new information or trying new behaviors.
- Use images, photos, diagrams, and graphics, since these formats reinforce the learning process by providing visual representations.

EXAMPLE: Many areas of Texas have been experiencing drought and significant economic and population growth. This combination strains water supplies, and consequently several communities are considering and implementing water reuse projects (see the San Antonio case in Appendix B). The Texas Section of the American Water Works Association, EPA, and the Water Environment Association of Texas developed the following revised water cycle (Figure 3-2) to promote public understanding and awareness. It provides a rich image, illustrated in simple, non-jargon style. Segments of the image can be viewed on smaller scale that make sense on their own, and when combined with other segments, illustrate the water cycle.

Information should be managed for the citizen volunteering time after a full day's work, the minister in the predominantly minority community, and the engineer or planner who is designing water reuse systems. If it fails to promote mutual understanding and opportunities to share perspectives among diverse participants, then it risks undermining public confidence and trust. The principle of managing information for all means that diverse individuals with varied informational needs and knowledge should have the venues and tools to listen to and learn from each other throughout the water reuse decision-making process.

Figure 3-2. The Complete Water Cycle.
Source: Texas Section of the American Water Works Association, EPA, and the Water Environment Association of Texas, 2001

3.3 Principle 2: Maintain Individual Motivation and Demonstrate Organizational Commitment

Individuals should find more good reasons (e.g., benefits for self and community) than bad reasons (e.g., real or perceived risks) to engage in public dialog. Organizations should demonstrate genuine commitment to engaging the public.

- Ensure and nurture multiple motives for the public to engage, including individual and community benefits.
- Demonstrate genuine commitment throughout the organization to hearing the public and its concerns.

Individual motivation and organizational commitment refer to the underlying reasons for engaging in a public discussion about water reuse.

- Why should people care about water reuse?
- Why should organizations care about what people think and care about?
- What should organizations do to encourage the public to care and in turn, to become involved?

Education and greater awareness and mutual understanding arising from *Principle 1: Managing information for all* may not result in sufficient motivation or organizational commitment to build and maintain public confidence. Do not make the mistake of assuming that educating the public about the issue will produce sufficient motivation or support to engender public confidence and trust.

The psychology literature has found very little support for the notion that educating the public, increasing awareness of relevant issues, or attempting to change attitudes or values will result in the intended behavior (e.g., acceptance of water reuse). Rather, individuals need knowledge and intent before they will act. Intent is shaped by many things, including motivation, interests and concerns, the convenience of the behavior, and perceptions.

Individuals need reasons to participate in water reuse decision making, whether they are motivated to participate on a citizen advisory panel or join an activist group that opposes a water reuse project. Rarely is a single motive enough. Usually, a person needs multiple motives to engage in a behavior. Sufficient reasons can range from self-interest (e.g., economic benefit, property values, reputations, personal enjoyment) to community interests (e.g., contribution to the community, opportunity for the community, hope for the future). Consequently, it is critical to communicate benefits, as well as respond to public interest in health and safety to ensure that sufficient motivations are identified to promote public participation. Likewise, broader policy discussions that connect water and wastewater management or the land-use planning discussion around growth and infrastructure have a greater likelihood of presenting multiple reasons for people to participate.

> **EXAMPLE:** n Gwinnett County, a water reuse initiative was expanded as a result of a broader discussion surrounding the update to the water and wastewater master plan. A multistakeholder advisory panel was formed to conduct the master plan update. The goals and objectives developed by the advisory panel included:

- Improve reliability of water and sewer system
- Develop strong maintenance and rehabilitation programs
- Protect public health and the environment
- Plan for water and collection system capacity proactively
- Minimize the negative impact of new facilities on neighborhoods and the environment
- Develop alternate water sources
- Pursue regional opportunities
- Manage water and wastewater demand
- Provide a high level of service at an optimum cost

The advisory panel evaluated all water supply and wastewater alternatives to indicate their preference in the near-term (present to 2010), mid-term (2010 to 2030), and long-term (2030 to 2050). While the wastewater preferences were generally more disperse among the advisory panel members than their preferences for water supply, nonetheless the advisory panel made recommendations regarding wastewater:

- Given the quality of treated wastewater effluent from the NAWRF, nonpotable reuse should be "pursued vigorously" through all periods (Hartley, 2001).
- Continue to seek conversions from septic tanks to public wastewater treatment.
- Discharge into the Chattahoochee River in the near-term, with discharge into Lake Lanier as a second option.
- Increase preferences for blended reuse in reservoirs for the mid and long term.

The advisory panel members wrote and edited the master plan update document. In all, they had considerable control over the process and tremendous authority over the decisions and the final document. The water reuse components of the plan were important for their relationship to other elements in an overall water and wastewater strategy. This broadening of the issues provided many more incentives for public participation.

EXAMPLE: The opposite end of this continuum of motivations can be seen in the San Diego case. A significant motive for conducting a water reuse project here was the lawsuit with the EPA. Conducting a water reuse initiative was an element of the settlement. Furthermore, several additional motives were clear for the city's sponsoring agencies, including the concerns about discharging excess reclaimed water to the sea, the reliance on imported water, and the potential for water shortages. The opposition did not view these as adequate reasons to move forward with the project and opposed the project based upon their perceptions of health risks, inadequate consultation and involvement during the decision-making process, possible socioeconomic or racial bias in the decision outcome, and availability of lower cost alternatives.

Multiple motives existed in both Gwinnett County and San Diego. However the nature of those motives produced different types of engaged publics.

Commitment, as discussed here, is considered an organizational-level factor in a public dialog. The organizations involved should demonstrate a genuine commitment to public participation, to listening to the public, and to taking public concerns seriously. The sincerity contributes to trust and the perception of fairness. "Faking it" will often be detected, and the artificial expression of concern for the public's view may cause considerable damage to public confidence and trust.

> **EXAMPLE:** The American Chemistry Council's (ACC) Responsible Care initiative requires its member and partner companies to develop mechanisms for outreach with the communities in which they operate. One such mechanism is the Community Advisory Panel (CAP), which the ACC believes "improves the dialog between plants and local communities, while building mutual respect and trust" (www.cmahq.com, 2001). The ACC describes the CAP and its function in the following manner (emphasis added):
>
> **What's a CAP?** A CAP consists of a group of individuals who live near or around a chemical facility and who represent the fabric of their community. CAP members have made a commitment to meet with the management of the local plant on a regular basis to discuss issues of mutual interest. *It is a forum for open and honest dialog between citizens and plant management to break down stereotypes.* The CAP meets periodically to *discuss any issues of concern that community members may have about the facility*. A CAP is an independent body. It is not controlled by the plant or the chemical industry. However, it should be supported and encouraged by the plant. *A CAP is not a public relations program designed to improve the image of a facility or the chemical industry. Nor is it an activist group for the purpose of closing down the local plant.*
>
> **Who belongs to the CAP?** *All key audiences within a community* should be represented on the CAP. These audiences may include environmental groups, civic leaders, business leaders, homemakers, hourly workers, and individuals who represent key elements of a community such as clergy, health care providers, emergency responders and educators. *Special attention should be given to ensure that the panel reflects the diversity within the community, including minority groups.*

While water resource agencies should demonstrate a commitment to engage with environmental, community, and other interest groups, the opposite also is true. Environmental, community, and other interest groups should demonstrate a commitment to engage with water professionals about water reuse. The result will be a constructive public dialog. Water utilities need to recognize what motivates opponents, what their interests and concerns are, and actively work to address those concerns, even if the interests and issues seem beyond the scope of the water reuse issue. By doing so and by building on the motivation of individual citizens and commitment of their interest groups, agencies demonstrate genuine interest in the concerns of citizens and are more likely to have a constructive conversation.

> **EXAMPLE:** Between 1996 and 1998, Gwinnett County, Georgia, updated its water and wastewater master plan. A multistakeholder advisory panel led the effort to evaluate all water supply and wastewater alternatives, with

responsibilities for writing and editing the final document. The representatives of interest groups on the advisory panel were charged "to act as a spokesperson for the group, representing the group's interest. The representatives were responsible for conveying information from meetings to other interested members of the stakeholder group" (Hartley, 2001).

A citizen advisory board was formed in Gwinnett County to oversee the water reclamation plant that was recommended in the water and wastewater master plan update. Board members are expected to stay involved for at least 5 years, until the plant is completed.

Commitment should permeate an organization. While leaders can show the way, workers on the plant floor and behind the wheel of utility trucks should demonstrate the commitment in their interactions with the public too.

EXAMPLE: Recently in Orange County, California, outside of Los Angeles, two charismatic disc-jockeys on a morning radio station began joking about the county's "toilet-to-tap" water reclamation program. The groundwater replenishment system initiative actually will use reclaimed water for seawater barrier and groundwater supply. A young engineer who worked for the county and listened to the radio show during his commute to work heard the jokes. He called the radio station and talked to the disc jockey about the program, clarifying misconceptions and communicating the benefits to the state and region. He made a difference and reached an audience that did not often consider the water resource issues of California (Water Reuse Symposium, 2001).

While for a few individuals a single strong reason may be sufficient, for many there needs to be more than one reason to participate in a water reuse decision-making process. If more of the motives are perceived as benefits than risks, then the engagement and the public dialog may contribute to building public confidence and the relationship between the utility and the community. When the motives are predominantly negative (e.g., perceived threat to public health or environmental quality) then the nature of public participation and dialog may tend to harm public confidence and damage the public relationship. A clearly demonstrated, genuine organizational commitment may go a long way toward producing more benefits to engagement, while an insincere commitment could quickly establish negative motives and damage public confidence.

3.4 Principle 3: Promote Communication and Public Dialog

Communication and the broader, cumulative public dialog should take multiple forms or venues and be on-going at all stages of a decision-making process.

- Expect to have to say the message multiple times (and in multiple ways) before it is truly understood by others.
- Expect to have to listen to others' messages multiple times and in multiple ways before truly understanding them.
- The quality of the public dialog is an indicator of the utility's relationship with the public and the public's confidence in the utility.

Discussion, the exchange of views, and other communication activities could be considered a powerful conduit through which all forms of information flow. Communication guides learning, shapes mutual understanding, and defines relationships among parties in a public dialog. Communication scientists have said that different groups or types of participants (e.g., citizens, elected officials, government managers, experts) communicate as if they speak different languages. Each group may articulate its interests, issues, and problem-solving perspectives in terms that may not be readily understood by other participants. Coupled with the nature of the complex information involved in water reuse and other environmental decision making and people's diverse learning styles, information often has to be communicated in multiple ways and repeated through several communicative channels and methods. Likewise, an individual should expect to have to hear a message several times and in different forms from other participants before understanding it.

In public participation processes that depend upon long-term, fairly routine communication among representatives from interest groups (e.g., advisory committees), the people who make up the constituencies of those representatives may fail to achieve the same level of mutual understanding, trust, and confidence that develop among representatives. Consequently, communication should include regular interactions and outreach to the representatives' constituencies.

Finally, highly effective communicators demonstrate many skills, including listening, seeking clarification, providing feedback, and modeling communicative behaviors. The critical role that communication plays in technical engineering professions has been well known for years.

> **EXAMPLE:** In a survey conducted 10 years ago by the American Society for Engineering Education, 4000 engineers were asked which academic subject area was most important for a successful engineering career. Not surprisingly many of the subject areas had to do with communication and interacting with the public and other individuals (emphasis added):
>
> - Management practices
> - Technical writing
> - Probability and statistics
> - *Public speaking*
> - Creative thinking
> - *Working with individuals*
> - *Working with groups*
> - Speed reading
> - *Talking with people*
> - Business practices (marketing, finance, economics) (Huckin and Olsen, 1991)

Public dialog is a broader concept related to the quality of the overall public discussion that arises from the culmination of individual communication events. While the effectiveness of the communication seems to be critical to the success of information exchange and mutual understanding, it is the *quality* of the dialog that appears to contribute to establishing the public relationship and confidence.

EXAMPLE: When people assess a speaker within the broader context of a public dialog, they are listening to the presentation and considering the speaker's:

- Sincerity and honesty
- Legitimacy and credibility
- Comprehensiveness, clarity, and coherence (e.g., no jargon or technical language)
- Accuracy (King, Felty, and Susel, 1998).

Many of these factors relate to the public relationship and confidence that exists between the receiver of the information and the speaker.

The extent of the public dialog may be a rough indicator of the level of public confidence in the utility and the health of the public relationship. While a hostile public dialog may appear to indicate a poor public relationship and serious problems, it may actually be a better sign than no public dialog. In other words, a family that knows how to fight and does so frequently might actually be more functional than one that avoids conflicts and does not work through problems.

The communicative principle plays a critical role in establishing public confidence and relationships. It transmits information, helps define motivations, and delineates social roles and expectations. In doing so, it carves the social landscape that defines whether a decision and the way in which people make the decision is considered fair and sound. It determines whether trust is built and maintained or is undermined and eroded. Consequently, it is critical to consider *all* of the messages and impacts that communications produce in a water reuse project.

3.5 Principle 4: Ensure Fair and Sound Decision-Making and Decisions

Participants should perceive the decision-making process and the outcome as fair and sound.

- Fairness includes process or procedural dimensions and outcome or distributive elements.
- Sound decisions are reasoned, well thought-out, and based upon accepted knowledge.

Fairness applies to both the decision-making process and the outcome.

- Was the process of making the decision fair?
- Was it reasoned?
- Did it involve everyone?
- Did participants listen to everyone?
- Did it address all concerns appropriately?

Fairness also relates to the outcome. That is, are the burdens and/or benefits shared fairly? A fair outcome may not need to be an equally distributed outcome. Hazardous waste facility siting cases have shown that some people will take on more burdens than others (e.g., one neighborhood will host the landfill), as long as the process of making the decision was perceived as fair *and* that there is some level of burden sharing (Rabe, 1994). In other words, while one community takes the landfill, the community sees that a neighboring community is committing to measurable, verifiable waste reduction activities, and a third community is hosting a waste transfer station. Everyone is seen as doing their part.

> **EXAMPLE:** San Antonio, Texas, initiated a nonpotable water reuse program in 1996, involving $125 million in pipeline infrastructure and 35,000 acre-ft/year of water for commercial, industrial, and agricultural customers. San Antonio has been experiencing drought and rapid population and economic growth that has strained water supplies and drawn down on the Edwards Aquifer faster than it can be replenished by scarce rainfall. In commenting about the water reuse project, city officials consistently stated (and were quoted in the local news media) that the project would "save aquifer water… enough to satisfy annual water needs of up to 80,000 households…[which is a] cost benefit to everyone" (Hartley, 2001).
>
> During the construction phase while crews were laying pipes and excavating roads and yards, an affluent neighborhood requested to tap into the discharge pipes for water for irrigation. The residents reasoned that if their streets were going to be disrupted to lay the main pipe, access pipes ought to be laid as well to provide residents a source of lawn water that would not be subject to drought restrictions. Upon consideration, San Antonio Water System managers denied the request, believing it was not fair to provide access to one community without providing a similar service to all residential communities along the two pipelines (Hartley, 2001).

A democratic expectation for fairness and justice exists. The public expects to have opportunities to state their interests, be heard, and be seriously considered in decision making. Furthermore, a little goes a long way, as Gwinnett County's approach demonstrates. Making a public commitment to acting fairly, designing a master planning process that was explicit about fairness, and taking a straightforward and honest approach with the news media resulted in credibility with the media. Likewise, in surveys conducted during water reuse initiatives and in other natural resource and environmental fields, the public consistently has expressed an interest in a sound, reasoned decision. However, the perception of what is sound and reasoned may vary.

3.6 Principle 5: Build and Maintain Trust

The cumulative attention to the previous principles contributes to the establishment and maintenance of public confidence and trust.

- Build trust and credibility reserves with the public and tap into them when necessary.

Trust is a comprehensive social concept with many components including belief and confidence in the reliability, credibility, and fairness of people, organizations, management systems, technologies, and other dimensions of decision making and implementation. Trust influences people differently, based in part on their cultural background and experience. In general, trust in government is in decline.

The public seems to trust and believe in science and technology for the most part, and they believe that adequate analysis is needed prior to making a decision. However, they may not always trust the information sources, including government, industry, and the media. The public seems to trust university scientists and the medical community more than other sources of information (Johnson and Scicchitano, 2000; Yearly, 2000). Nonetheless, adherence to expert knowledge and perspectives at the expense of and disregard for local knowledge (that is, expertise that local community members obtain from experience with their specific natural, physical, and social environment) and local concern leads to mistrust and suspicion within local communities, particularly when the expert knowledge is inconsistent with local knowledge or perspectives. In such cases of inconsistency between expert and local knowledge, the motives of the expert or scientist may be called into question, and their information may be discarded by the public.

Trust is built through communication and dialog. Trust has been shown to be enhanced in cases of high risk perception when:

- Dialog is sustained.
- The public has independent sources of information, not linked to the sponsoring agency.
- The public can ask questions.
- The public is involved early.
- Information is available to everyone.
- Behavior is non-coercive. It is considered a reasoned, fair way to make a decision.
- Everyone's opinion matters, and there is a willingness to listen to all views and expand the discussion if necessary.
- Citizens have some level of control in the process (e.g., contributing to the agenda or ground rules) (Renn, Webler, and Wiedemann, 1995).

EXAMPLE: Gwinnett County achieved a high level of trust with many constituents, including the media, from their approach to communication and the overall attitude of county officials. Gwinnett County has built a $245 million, 20 mgd water reclamation facility, with a $330 million, 40 mgd expansion underway. During the development of a water and wastewater master plan from 1996 to 1998, the county Department of Public Utilities formed an advisory panel comprised of multiple stakeholders (e.g., homeowners, businesses, developers, water users, county city officials, environmental organizations, and citizens). The advisory panel members wrote and edited the master plan. County officials facilitated the meetings to promote active involvement of all interests despite the ongoing challenge to "maintain balance among participants and prevent some individuals from dominating" and to remain "neutral" (Hartley, 2001). While a professional facilitator may have helped address these challenges, the county officials felt it was important "not to put hired help between us and the citizens…If we wanted to build trust, [obtain] a higher degree of credibility, then we could not hide [behind] a professional facilitator (Hartley, 2001)."

EXAMPLE: Similarly in the case of a biosolids project, the Denver Metro Wastewater Reclamation District (Metro) learned from their experience with land application of 80 dry tons of biosolids/day on a Metro-owned 52,000 acre farm east of Denver. According to one Metro official, the "its my land so I can do whatever I want" approach failed, leading to considerable public opposition. To rebuild the trust, Metro committed to an approach that demonstrated attention to the concerns of neighbors, farmers, local political leaders, and soil conservation advisors. They took a thoughtful, open, honest, and patient communication approach that included frank discussions about risk. In addition, Metro funded an independent monitoring program to ensure that "good, trusted" information was available (WEFTEC, 2001).

Think of trust as a good friend and neighbor. Friends are not made when needed, but if a friendship has been established with a neighbor, that neighbor will be available when needed. A former public health official said that public officials should, "get to know [the media] during times of no crisis so you have credibility when a crisis occurs."

3.7 Conclusion

The five principles are not independent of one another, rather they interact and reinforce each other. Trust, for example, grows through effective communication, but the presence of trust promotes effective communication. The complex interaction among social science factors has made it difficult to clearly identify causal relationships and theoretical connections. Nonetheless, analogies may facilitate appreciation of the power and interaction among these principles. While the analogies may imply causal relationships or correlation between the principles, they should be considered in light of the image they produce in the readers' minds and not as theoretical implications.

First, the five principles can be thought of as the components of photosynthesis and the growth of a tree. Fundamental components of the photosynthetic process, such as light from the sun, are analogous to the basic pieces of information and the underlying individual motivation and organizational commitment upon which a tree and *public confidence* grows. A remarkable and powerful transformation of these basic components takes place as photosynthesis converts light to plant energy in the form of sugars, just as the communicative process transforms all types of information and the motivation and commitment into a *constructive public dialog*. The energy is used to build cells, and block-by-block it forms the tree's fibrous pulp. Likewise, communicative acts and a constructive public dialog build the *perceptions of fairness and the soundness of decision making*. Finally, the plant's tissue is organized into a large, solid, and sturdy tree. Communication and fair sound decision making builds a *robust public trust*.

Light: | Photosynthesis converting light into useable energey: | Energy to build plant tissue that form a solid, sturdy tree:

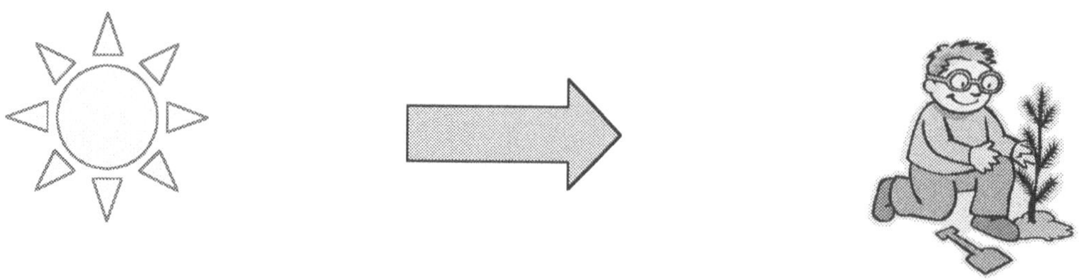

Information
Motivation and commitment | Communication and public dialog | Perceived as fair and sound
Trust

Figure 3-3. Growing a Solid Tree of Public Confidence.

Another way to look at the five principles is as a river carving the landscape over geologic time. In this analogy, communication is water that carries with it all the forms and types of *information and the individual incentives or motives and organizational commitment*. The individual molecules of water by themselves are significant in the same manner as information, motivation, and commitment are important, although it is their cumulative impact when they come together in a river (i.e., *the public dialog*) where their power is the greatest. The water flows through a sometimes meandering and sometimes raging course of public dialog in a *decision-making process*. As it flows and over time, the communication and public dialog *carve the social landscape, defining what is fair and sound and establishing or eroding trust*.

Both of these analogies serve two purposes. The first is to create vivid cognitive images that illustrate the rich relationships and influence of the five social principles at play in public perception and participation. The ability to grasp and make connections among these principles in the reader's mind is determined by the reader's existing knowledge structure and experiences. The second purpose is to provide repetition of the message, so that the reader hears it again and again, and possibly hears it in a manner that is most consistent with their individual learning style.

CHAPTER 4.0

PUTTING THE PRINCIPLES INTO ACTION: DIAGNOSTIC TOOLS TO HELP TAILOR PUBLIC OUTREACH, EDUCATION, MARKETING AND PARTICIPATION PRODUCTS AND ACTIVITIES

Once more, the principles are not a checklist of "to-dos" that establish public confidence. No one should suggest that in any water reuse project, success will be achieved by doing X, Y, and Z at time periods A, B, and C. Quite the opposite, the research suggests that a one-size-fits-all approach cannot work because the most appropriate and effective tools and techniques vary from case to case. Rather, what is critical is the utility's commitment to the tailored application of the principles based upon the context of the community, the capabilities of the participants, and the decision at hand. But how can the application of the principles be tailored?

Two suggestions are put forth in this chapter. First, a set of diagnostic questions and data analysis strategies can help water professionals assess the context in which a water reuse decision takes place. With a more thorough understanding of the context, monitored on a regular basis to detect subtle but significant changes, water professionals can better prepare to select specific public participation, outreach, education and marketing activities. Second, data collected and analyzed then need to be interpreted to inform the necessary site-specific modifications of the public outreach, education, participation and marketing tools to achieve the framework principles.

4.1 What Are Diagnostic Tools?

Diagnostic tools are questions that can help wastewater professionals construct a comprehensive community profile. The questions should be asked of many community participants and stakeholders and measured with multiple methods, from surveys, interviews, and focus groups to media content analysis and public meeting observations and video analysis. Data analysis techniques should identify patterns and significant differences that characterize the context and challenges water professionals face. Once a profile is constructed and analyzed, on-going validation should take place to continually confirm the context and detect any subtle but potentially significant changes. Painting the picture of the complex social structure of a diverse community and monitoring the interests and values within groups in the community on a regular basis helps ensure public outreach, education, participation, and marketing activities can be continuously tailored to a community. Water reuse initiatives should institute an ongoing diagnostic program.

Below are some suggested questions that serve as a starting point for exploring the community at the societal, institutional and organizational, and group and individual levels. Asking these questions will likely lead to several additional follow-up questions that are case specific but critical for producing a comprehensive picture of the community.

4.2 Societal and Institutional Context

Broad societal factors that relate to the underlying power structure and historical and cultural factors are the foundation upon which any public decision takes place, including water reuse projects. As a starting point to map this complex social and institutional structure, consider the readily available descriptive statistics of a community (e.g., the basic demographics of the community, including any changes and shifts in the demographic make-up of a community over time). Changes in the demographic make-up (e.g., racial, ethnic, or socioeconomic composition of a community) may help identify potential underlying conflicts between different groups. These conflicts may have erupted into public disputes and, perhaps, serious tensions in the past or might simmer close to the surface.

However, the basic, underlying demographics are only the first layer of a community profile. Within this demographic landscape is a communication network and power structure. Communities possess social capital, that is the set of behavioral norms, organizational networks, and social trust among community members (Putnam, 1995). Community leaders exist within and among all of the demographic groups and they wield a tremendous influence on community opinion, interests, and motivations. They are critical nodes of power and influence in the social network. A water professional should map the networks, since a network map may provide insights into how best to craft particular messages and information and communicate this information (e.g., what opinion leaders to target for outreach materials, what communication channels are most trusted and used).

Be aware that many groups may have an interest that they are not yet aware of. Just because a group isn't speaking up does not mean that they do not have a concern or interest in the water reuse issue. Some may have interests in the issue, but have yet to organized themselves into a distinct group. It would be better to reach out to members of these "unorganized groups" and even assist them in organizing before they mobilize around a perceived threat or risk from a water reuse project. In other words, be proactive and build the utility into the fabric of a community's social capital, thus building the public trust and confidence before it is needed during a particular water initiative.

Institutional procedures, such as legislative decision-making authority, budgetary appropriations, judicial oversight, and other broad structural democratic characteristics define the constraints and opportunities for public participation. The institutional structure also contributes to defining the level and type of engagement that can most readily occur between the public official and the citizens. It contributes to the nature of the norms, networks, and trust that underlie social capital.

Likewise, organizational cultures establish powerful behavioral patterns, norms, and expectations that influence overall social capital. Organizations take many forms, including less formal organizational units, such as citizen groups, block groups, or social networks, as well as large and small government agencies and corporations. In all organizations powerful behavioral norms can result from unique cultures.

Several questions can be asked to help paint the big picture and survey society's power, historical, cultural, and institutional context (see examples in Table 4-1). Different techniques can be used to obtain answers to these questions, including surveys, content analysis of past and present media coverage, focus groups, and interviews. Multiple information gathering tools should be employed and integrated into a comprehensive community assessment of the diversity of perspectives.

Review organizational policies, standard operating procedures, and organizational structure charts for the water utility and other organizations involved in the issue to understand institutional expectations. Also interview legal and policy staff within the water utility and community leaders outside of the organization who may hold their own perceptions of the societal and institutional structure. Talk to community leaders in non-profit organizations, churches, block clubs, and other groups. Ask these leaders who else should be contacted. Be proactive and reach out to as many of the diverse community groups as possible, including those groups that may not immediately express interest in a water reuse project.

Note that it is difficult to identify an organization's culture through direct questions. People who are functioning under certain norms are not always able to point to the particular norms that are contributing to why they are doing what they are doing. However, using multiple methods such as combining interviews and surveys with document content analysis of organizational documents and media coverage can begin to identify powerful institutional and cultural factors. Organizational psychology, public policy, and business departments of local universities often have experts in cultural factors who can assist developing additional measures.

Table 4-1. Diagnostic Questions: Societal and Institutional Context
Descriptive Profile: Collect demographic data from the U.S. census (online), local planning departments, or other sources. ♦ Identify different cultural, ethnic, socioeconomic or other segments of the community. ♦ Where are they located? ♦ Look at past demographic data to see trends or shifts in community make-up.
Community Network Profile: Conduct interviews and/or surveys of residents, review local newspapers and community groups' newsletters, and other sources. ♦ Who are group opinion leaders and community leaders (e.g., community-based organizations and non-profits)? ♦ What are the communication channels or networks used by various groups within the community? How is information shared and learned? ♦ What coalitions or associations of community groups exist? What are their shared interests? What are their disparate interests? ♦ Would any of these groups have a stake or an interest in the water reuse issue?
Community History Profile: Interview and/or survey long-time residents and leaders and review past issues of local newspapers and community groups' newsletters and other sources. ♦ What are the prior civic activities or experiences of each group? What issues are salient to them and why?

Table 4-1. Diagnostic Questions: Societal and Institutional Context (cont.)
Community History Profile (cont.): ♦ Have disputes emerged in the past in land use, facility siting, or other community planning activity? Has the public debate included underlying cultural, ethnic, or socioeconomic differences? ♦ Do various groups have different perspectives on what is considered fair or whom is a trusted source of information? ♦ Are there historical relationships between community organizations (e.g., alliances or rivalries)? Are there traditional societal norms of behavior or communication between groups?
Organizational Profile: Use multiple methods of collecting information, including interviews, surveys, document reviews, and focus groups. Consult with local university social science departments. ♦ Are there multiple organizational cultures? In what ways do the cultures complement or conflict with one another? ♦ Are organizations experienced with multiple public outreach, education, and participation activities? If so, how? If not, why not? ♦ Are organizations experienced coordinating with each other? If so, how? What are the challenges and barriers to coordination? ♦ Do organizations have the capacity to conduct public outreach, education and participation activities with their own constituencies?
Institutional Profile: Review laws, regulations, policies, and guidances. Interview policy and legal staff and outside leaders and stakeholder representatives. ♦ Who is the decision maker and what public participation activities and decision-making processes are mandatory? ♦ What is the extent of decision-making authority? What can be decided, and what is beyond the scope of the issue? What decisions are anticipated and when? Who has the authority to make decisions? ♦ Are there various institutional procedures interacting with one another (e.g., litigation, multiparty negotiations, advisory boards, expert panels, and others)? How might those institutional interactions be contributing to public perceptions of the issues? ♦ Given the societal context, are the required participatory activities likely to contribute positively or negatively to developing public confidence and trust?

4.3 Group Profile and the Decision Context

Within these societal and institutional structures there are powerful group dynamics that play out as people come together in prescribed decision-making processes. Several factors contribute to the group dynamics within processes, including those manageable issues that contribute to public perception and the nature of participation (e.g., communicating complex information, addressing perceptions of fairness, managing representation of interests, and designing decision-making processes). Furthermore, a person's academic discipline, professional training, or experience can establish strong behavioral norms, expectations, and values. It can be challenging to effectively communicate and generate mutual understanding across groups that are well trained and have considerable experience. These factors may contribute to the overall quality of the public dialogue and the degree of empowerment experienced by members and groups within the community.

Diagnostic questions would aim to provide an indication of the group and decision-making context. Different techniques can be used to characterize the group dynamics and processes under which water reuse decisions are made, including reviewing organizational policies, standard operating procedures and organizational structure charts for the water utility and other organizations involved in the issue. Interview legal and policy staff within the water utility as well as community leaders outside of the utility who may hold their own perceptions of the group profile and the decision-making process. Likewise surveys, content analysis of past and present media coverage, and focus groups can probe group and decision-making factors.

Again, note that it is difficult to identify professional culture factors through direct questions. People who are functioning under certain norms are not always able to point to the particular norms that are contributing to why they are doing what they are doing. However, using multiple methods such as combining interviews and surveys with document content analysis of organizational documents and media coverage can begin to identify powerful institutional and cultural factors. Organizational psychology, public policy, and business departments of local universities often have experts in cultural factors who can assist in the development of additional measures.

Many other issues important in group and decision-making processes are captured under other contextual categories (e.g., interests, values, and concerns discussed under in the section below on an individual's information processing). Nonetheless, multiple information gathering tools should be employed and integrated into a comprehensive assessment. See Table 4-2 for a few starting questions.

Table 4-2. Diagnostic Questions: Group Profile and Decision Context
Group Profile: Review media coverage and interest group publications and community organizations' newsletters. Interview community leaders, stakeholder group representatives, and citizens. ♦ How many parties are there with an interest in the issue? ♦ What stake do each of the parties hold in any given outcome? ♦ What opportunities exist for the parties to express their interest or get their interest met? Are parties feeling empowered or disempowered by the process? ♦ What information or facts do parties agree upon? What information or facts do parties disagree upon? What information gaps exist? How are parties framing the issues? Do parties have different access to information? ♦ Do participants think the process is fair? ♦ Are public interactions among the interested parties, including the behaviors of individuals, perceived by others as destructive to the decision-making process or the group?
Professional Profile: Use multiple methods to collect information, including interviews, surveys, document reviews, and focus groups. Consult with local university social science departments. ♦ What are the academic, professional, and/or experiential background of representatives and leaders in organizations/groups? What are the academic, professional, and/or experiential background of the journalists covering the water issues? ♦ Do individuals within these categories of expertise have experience working across disciplines?

4.4 Individual Characteristics

Individuals have strong psychological biases that promote thinking and functioning in a complex world, but also impact how they learn, their perceptions and reactions to risks, and their judgment. Furthermore, individuals bring specific skill sets and capacities to their interactions on a public issue and in public settings, along with their power, leadership, and behaviors that accompany their positions and personalities. Individuals also come with attitudes, risk perceptions, learning styles, and values.

Using diagnostic questions to profile the community of individuals is nearly impossible because there will be too many people to reach. Rather, it will serve the water professional better to gain an appreciation for the degree of variation within the community along a number of professional and individual contextual characteristics. It will be important to sample perspectives on these questions from as many groups within a community as is practical, ensuring that the cultural, ethnic, socioeconomic variation is accounted for in the collected information. Do not discount the anecdotal stories and opportunities to interact with citizens and the others from the public in informal settings (e.g., coffee shops, local sports events, block club parties) and less structured interactions during formal public participation activities (e.g., at the sign-in table or during a break). Furthermore, more in-depth, targeted analyses of opinion leaders and those that shape public opinion are worth the effort. See Table 4-3 for a few suggested starting questions.

Table 4-3. Diagnostic Questions: Individual Context
Individual Profile: Sample broadly to understand the range of individual characteristics in the community. Use multiple methods, including interviews, surveys, document review, media coverage review, and other data sources. ♦ What are the different interests, values, and concerns of various groups and their leaders? What do they care about? In what areas do they differ in what they care about? What interests or values they hold? ♦ What are the leadership styles and personalities of representatives and leaders? ♦ What are the negotiation, communication, and analytical skills of representatives? ♦ What are the learning styles of the representatives and the general public, particularly those attending public meetings? ♦ In what way might the individual and organizational capacity differences among people and parties produce power imbalances or other advantages and disadvantages among participants? ♦ What risk perceptions and attitudes (e.g., degree of risk aversion) exist among representatives and the general public? What do members of the general public care about? What are their values and underlying concerns? ♦ What is the capacity of representatives or stakeholder leaders and constituencies to manage and understand complex scientific or technical information? ♦ Do individuals exhibit an "expertism," or some degree of extreme confidence that their knowledge and disciplinary perspective is all that is needed to solve a problem? If so, what impact is this having? ♦ Do decision makers look like the community? In other words, is the demographic make-up of representatives participating in the decision-making process similar to the community? If not, what concerns might this difference raise and with whom? ♦ Probe the water knowledge and awareness levels of citizens.

A comprehensive community profile based upon a diagnostic assessment of societal, institutional, group, and individual level factors can prove invaluable in providing a picture of what the utility has to work with. If specific factors within one or more of these categories prove particularly important, examine the attached literature review on that topic more closely to see additional dimensions worthy of analysis. Targeted, in-depth analysis of contextual factors that are particularly relevant in a specific case could be very important to improving understanding of the community and in turn contribute to tailoring the public outreach, education, participation, and marketing programs.

Once this landscape is defined, continue to monitor it to detect changes. The utility will be better prepared to tailor public outreach, education, and participation activities to the specific context. The diagnostic tools are most effective if the analysis includes multiple techniques of capturing the answers to the questions discussed above and if the tools can be employed continuously. Confirming the community profile and detecting changes can be critical to ensuring long-term effectiveness of a public outreach, education, marketing, and participation effort. Employing mismatched outreach, education and participation activities may be, at the least, ineffective and a waste of resources, and in the worse case, harmful to public confidence and trust. For example, using formal training workshops with Ph.D. presenters for a socioeconomically depressed group may be perceived as paternalistic, disrespectful, and belittling of local values and concerns. This could lead to greater suspicion of motives and distrust.

4.5 What Should be Done Now? Analyzing Community Assessment Data

While information gathering is crucial to gaining an understanding of the community, analysis and interpretation of that information key to tailoring public outreach, education, participation and marketing efforts. Analysis identifies patterns and changes in factors that may impact and influence how the framework principles apply in a particular community. Some analytical techniques entail advanced statistical models, while others can be sophisticated approaches to finding patterns in qualitative data, such as mapping and other graphic strategies to see relationships and correlation among coded qualitative concepts. Local universities and colleges can be a resource for more sophisticated analytical techniques. Nonetheless, fairly straightforward counting and simple manipulation of data can identify patterns and anyone can apply these techniques.

However, fairly simple descriptive statistics and the discovery of key themes and patterns within certain segments of the community can be tremendously helpful in guiding a tailored application of public outreach, education, marketing, and participation programs. Likewise, identifying changes and shifts in themes or concepts can be important clues to issues that may need to be addressed within a water reuse decision-making process. The analysis does not need to be overly sophisticated, but the data collected need to be critically examined. Collect this information on an ongoing basis, so that changes in the patterns can be identified along with verification of whether the utility's efforts are affecting the community.

From a practical standpoint, it is more constructive to focus on the public outreach, education, and participation activities that can be influenced or controlled. Keep in mind that the community profiles will assess factors well beyond the utility's control. Do not expect to change all the community assessment measures significantly. Rather, look at them as guides to a tailored application of programs, consistent with the framework principles. The framework recommends a "principle-based" management strategy, that is public outreach, education, and participation programs that are guided by a comprehensive and ongoing community diagnostic assessment that aims to satisfy the framework principles.

> **EXAMPLE:** The demographics of a hypothetical community may include a substantial African-American population. However, in recent years the Latino population has surpassed the African-American population. Like the African-American population, the Latino population is diverse, comprised of people who have a Mexican, Central American, or South American heritage. The community also features distinct Korean, Chinese, and Vietnamese neighborhoods. The community also contains distinct religious organizations, with some paralleling racial and ethnic lines more closely than others. The tensions that have flared up in the past between African Americans and Caucasians are receiving less attention as new dynamics get more media coverage with the increasing diversity of the community. Media outlets are publishing and reporting more information in Spanish than they were 5 years ago. Few of the established African-American organizations and none of the Latino groups have expressed any interest in water reuse, as demonstrated by attendance at public meetings and opinion surveys. The major leaders from all the minority and religious groups have remained silent on water issues. At the same time, leaders in the African-American churches have been vocal in land use, urban revitalization, police behavior, and other city issues and are actively engaged with the city council and city departments. Latino leaders have emerged in the business community, with particular interests in community development and revitalization.
>
> The community assessment shows a utility with a strong professional norm and culture dominated by experienced environmental engineers who have developed innovative technical solutions to most of the problems faced over the past 20 years. In the past 5 years, there has been increasing staff turnover, with several senior engineers retiring and recent graduates joining the staff. Stories recounted by senior engineers paint a picture of an organization that has increased its interest and efforts to engage the public about water reuse, in part to improve the public trust and confidence in the utility. However, substantial organizational resistance remains and budgets are minimal for public outreach, education, and participation. The wastewater and drinking water departments have been perceived as rivals by many for years, only reluctantly coordinating with each other and often at outsiders' insistence (e.g., elected officials). The drinking water department has the larger public outreach, education, and participation budget but often communicates messages about wastewater that are not supportive of some water reuse applications.

From this simplified community assessment example, several patterns emerge that can influence a public outreach, education, marketing, and participation strategy. More detailed information from a community assessment can provide guidance on specific messages and activities that would be most constructive. Nonetheless, here are some general approaches based upon the above scenario:

Managing information for all. The different groups may have varying degrees of experience with engaging with public agencies over a public decision. Information supplied to them should be in their language and may need to introduce how water decisions are made, and how water is managed in the community, information about the water reuse decision. It may be best for the learning of the overall community if information is presented incrementally through a series of educational meetings, workshops, and informal settings.

Some journalists in the community are reporting in Spanish, and other community group newsletters and publications are printed in other languages. Reaching out to these information outlets and their reporters and staff might help get the word out. These journalists and the leaders and staff of community groups might also have suggestions for ways to best reach the diverse constituencies. It is important to reach them all. It is likely that different outreach, education, marketing, and participation activities will be needed for different cultural, ethnic, and religious groups.

Maintain individual motivation, and demonstrate organizational commitment. While some groups have not engaged in water issues before, it is important to reach out to the leaders of these groups to ask questions. How are they impacted by water? What do they care about and why? Connections between their interests in urban revitalization or community development and water that they have not yet identified may be observed. It may seem counter-intuitive, but it is better to "organize" these groups and help their leaders see their interests in water before they "discover" the connections themselves later, possibly through polarized media coverage. It is likely that different groups will have multiple reasons for wanting to engage in water reuse discussions. It will be important to continue to keep these diverse motivations in mind and continually "tailor" outreach, education, marketing, and participation activities accordingly.

Ensuring that organizational commitment to public engagement is demonstrated throughout outreach, education, marketing, and participation materials will be more challenging given the organizational resistance within the wastewater department and the rivalry between the wastewater and water departments. It will be important to select a technical spokesperson carefully. An internal effort to reach out to the water department and its public involvement and communication staff would be worthwhile, as competing messages about water reuse from two departments within a community would confuse the public.

Promote communication and public dialogue. The increasing diversity of the community introduces the possibility that underlying tensions and competition between and among minority groups and the majority Caucasian population may emerge in any public forum, including the water reuse decision-making process. At key public meetings it may be useful to have staff or an outside professional with facilitation and mediation skills available to guide and manage underlying conflicts in a constructive manner. Translators may be needed at meetings. Using multiple meeting formats may be most conducive to engaging all segments of the community. In fact, extra effort may be needed to ensure that advisory committees or blue ribbon panels reflect the diversity of the community.

The diversity of the community and the language barriers that may exist means that it will be even more important to design and conduct multiple "listening" sessions with the constituencies. It will not be easy to develop mutual understanding among the groups without an extensive amount of active listening on all parts.

Ensure fair and sound decision making and decisions. Fairness may be defined differently by the different socioeconomic, ethnic, and religious groups. Likewise, different segments of the community may weigh fairness in the decision-making process and outcome, and what makes for a sound, reasoned decision differently. Differences between how the professionals in the wastewater department view fairness and sound decision making may also be observed. As described in this scenario, the senior engineers and those who show resistance to increased public engagement may weigh scientific and technical soundness very high and of greater importance than the public perception of fairness and soundness. Awareness of these differences ahead of time can be helpful.

Build and maintain trust. Reach out to all the diverse community groups. Trust is built and maintained over time. Demonstrating that the utility is doing its best to meet the high standards outlined in this framework will go a long way toward building trust in diverse communities.

In the end, it is what is done with the community data and assessment that is critical, not the assessment itself. An assessment can be used to guide decisions and tailor outreach, education, marketing, and participation products and activities. In this way, it can help the utility build and maintain public trust and confidence.

CHAPTER 5.0

PUTTING THE PRINCIPLES INTO ACTION: TIPS FROM "PEOPLE" SPECIALISTS

The framework was developed through a year-long, multidisciplinary exercise. Several experts and experienced water professionals reviewed background materials, wrote white papers, and met to consider and discuss public perception and participation in water reuse. This chapter recounts those perspectives. Commentary includes discussion on some public outreach, education, and participation strategies, tools, and techniques that have been tried, tested, and refined in the field. Some of the tips and lessons have been learned the hard way, while others have resulted in real progress on challenging water and other natural resource issues.

This chapter provides a range of tips supplied by experts and experienced water professionals from many fields, including:

- Consensus-building and dispute resolution
- Environmental engineering
- Change management
- Public policy
- Cognitive psychology
- Risk communication

This chapter includes tips from the following experts:

- Maureen Austin of the University of Michigan
- Gail Bingham, Juliana Birkoff, and Troy Hartley of RESOLVE Inc.
- James Conklin of Information Development and Training Inc.
- Mary McDaniel of McDaniel Lambert Inc.
- Michael McGuire and Edward Means of McGuire Environmental Consultants

After instituting a diagnostic assessment program, consider this chapter a starting point for selecting and tailoring public outreach, education, and participation activities.

5.1 Consensus-Building Processes – Gail Bingham, Juliana Birkhoff, Troy Hartley

Dispute resolution and mediation techniques have been increasingly used in natural resource and environmental management decision making over the past few decades. Several lessons that focus on the characteristics of the decision-making process have emerged. Consensus-building processes are not always appropriate but are constructive when:

- There are several ways the issues can be addressed or resolved.
- The process will not cause unreasonable delay.
- A means of implementation exists for agreements, if reached.

Under these circumstances, water professionals should consider several design and process issues when using a consensus-building approach. First, participants should agree on the *terms for the process*. For example, all interested parties should be involved and reach agreement on the objectives for the consensus process. This includes building an agenda that all parties find acceptable, establishing realistic deadlines, and planning for implementation. If a third-party facilitator is used, parties should agree upon the facilitator's role. Ground rules should consider issues of whether observers are allowed, how alternates for representatives are identified and used, and the procedures for adding parties if appropriate. Furthermore, ground rules should identify acceptable procedures for calling for side-bar caucus meetings, changing the agenda, and interactions with the media. While not a ground rule, asking questions and effective listening are behavioral norms that should be encouraged among all participants.

Important *interorganizational* factors also need to be considered. It is critical that all representatives keep their constituency informed on a consistent basis. Water professionals who use consensus-building processes should understand their own and each party's internal decision-making process. The outcomes of consensus processes often require confirmation through internal decision-making of an organization. Participants should understand the degree of freedom that each negotiator has been given by their organization or stakeholder group. An internal team communication plan is prudent to ensure timely information sharing. The negotiating group should also think ahead and plan for what will be necessary to obtain ratification of an agreement.

Facilitators take many of the negotiation management and transaction costs from participants so that participants can focus on the issues. Facilitators can help form and maintain coalitions of interests and help parties shape processes collaboratively, including jointly developing agendas. Facilitators identify subsets of issues for subcommittee work, and plan time for and assist in draft revisions. Furthermore, third-party neutral facilitators assist in clarifying team roles and establishing efficient mechanisms for communication in and between meetings.

Water reuse and other natural resource issues can present *multiple issues, technical complexity*, and *scientific uncertainty*, all of which present management challenges in consensus-building processes. Through years of experience, several strategies have emerged to address these complex informational issues by identifying potential linkages and devising strategies for jointly collecting and analyzing data (Table 5-1).

Table 5-1. Addressing Complex Issues in Consensus-Building Processes
Seek Linkages
♦ Identify priorities of parties
♦ Identify subsets of issues for subcommittee work
♦ Avoid reaching closure too soon on a few issues
♦ Plan sufficient time between meetings
♦ Agree on procedures for handling unresolved issues
Conduct Joint Fact Finding and Analysis
♦ Identify what is and is not known
♦ Identify what facts are in dispute
♦ Define problems before seeking solutions
♦ Help parties articulate assumptions
♦ Create a system for sharing information
♦ Seek mutually acceptable criteria for decision making

There are tips to address the *power and resource disparities* that always exist among parties. The disparities are not always readily identifiable, since power and resources can be obtained from multiple sources, including alliances of parties, information, precedents, and solutions to others' problems. It is beneficial to help weaker parties obtain sufficient resources and help them negotiate more effectively. While it may seem counter intuitive, it is often in the best interest of all parties to strengthen weaker or organize unorganized interest groups so that they are more capable of negotiating effectively. Implementation of an agreement that inadequately addresses the interests of an unorganized interest group can be undermined when the constituency of the unorganized interest group sees the agreement. Their interest will be sharpened when they see that they really do care about the outcome and it fails to protect what they care about (Birkhoff, 2001; Bingham, 2001).

In addition mediation research and theories have developed around the concept of stages in consensus building. Bingham (1998) has discussed three stages: convening, substantive dialog, and implementation, each with specific goals and objectives.

The initial convening stage aims to reach agreement among parties on the purpose, products, and process of the dialog. It achieves this through interviews among stakeholders and by designing a decision-making process for review and comment among stakeholders.

The substantive dialog stage involves an opening, middle, and closure of the dialog. The goal of the opening is to reach a shared understanding of the problem. It entails confirming ground rules, testing assumptions, defining problems, and developing mutual understanding of interests. Furthermore, the opening includes activities to clarify issues that must be addressed, developing criteria to evaluate alternative solutions, and identifying information needs and strategies for collecting and analyzing data. The aim of the middle phase of the substantive dialog is to explore possible outcomes through the development of creative options and evaluation of those options relative to the criteria. Finally, the closure of the substantive dialog reaches recommended solutions. Reaching closure involves narrowing options, group problem solving, and tradeoffs by stakeholders. To reach closure efforts will be made to consult with decision makers and other constituencies of the negotiating representatives, agreements will have to be drafted and revised, and planning for implementation of the agreement will begin.

The third stage, implementation, has an objective of observable change. Implementation can involve ratification procedures, monitoring progress, and meeting milestones. Occasionally further group problem solving and renegotiation may be needed, based upon the nature of implementation.

See Moore (1987) for a further discussion of stages. He builds a 12-stage process of achieving formal agreements through mediation. Please note that much of the development and research on the concept of stages has emerged from the dispute resolution literature, focusing on agreement-oriented processes to resolve specific issues at dispute. While many examples of specific disputes in water reuse cases where the agreement-focused strategy of stages would be applicable are available, the emphasis of the framework is on building public trust and confidence. In this context, it is less clear that the stages approach can be extrapolated to the long-term relationship level of public trust and confidence.

5.2 Leadership and Politics – Michael McGuire, Edward Means

Experienced environmental engineers and government managers who have lived through drinking water and water reuse conflicts have raised concerns about which agency should lead a water reuse initiative and what strategies should be used to interact with political leaders. First, some suggest that responsibility for countering the effective and catchy "toilet-to-tap" phrase should lie with the drinking water agency or some new entity that serves as the *lead government agency*. The message should be supported and carried by the chief executive officer of that agency. The message can be put in terms that people care about with an emphasis on the "green" water reuse issues (e.g., conserving limited fresh water supplies for higher order uses, maximizing a finite resource, taking something considered to be a waste and turning it into a useful commodity, and preventing or minimizing the discharge of wastes into the environment).

Likewise, the community's *political leadership* did not support many water reuse projects that stopped under public opposition. In addition to underscoring the need to reach out to all potentially interested groups and not leaving anyone on the side lines, it is also important to reach out to the community's political leaders through techniques for regularly consulting with political leaders. For example, a political advisory committee may complement existing citizens' advisory committees (McGuire and Means, 2001).

5.3 Change Management and How People Deal With Behavioral Change – James Conklin

Over the past few decades, a growing awareness has emerged that one of the key factors underlying the successful introduction of a new scientific or technical breakthrough is to take explicit steps to promote acceptance and use by the public. Initially, this led to the development of a variety of disciplines collectively referred to as change management. Change-readiness assessments were undertaken, communication strategies were devised, and the familiar axioms of the 1990s – "communicate, communicate, communicate" – were proclaimed in boardrooms and town halls.

The very term change management, however, contains the seed of its undoing. When dealing with science and technology and diverse stakeholders, it is misleading to suggest that social change can be *managed*. Applying proscriptive, top-down management techniques to force consensus and adoption will usually be seen for what they are – efforts to manipulate people into supporting a new initiative. Other strategies are more effective at dealing with change constructively.

First, an effective and meaningful notion of "participation" must *supplant the recent emphasis on manipulative techniques* to promote "change management" and "buy-in." Both theory and practice in organizational change suggest that effective participation involves the creation of trust and collaborative relationships among stakeholders. Participation is not limited to "providing input." Instead, it must clearly include the ability to shape and influence outcomes and to share in the decision-making process. Those responsible for future water reuse initiatives would be well advised to follow processes that permit full and effective participation by all stakeholders.

This means that effort must be made to help stakeholders overcome any knowledge barriers due to specialized technology or complex science. Those in authority must be prepared to share decision making. A basic model for participative action should include steps to create relationships, communicate and learn, create new understanding and knowledge, and move to action.

Second, *participation* must begin *early, and be continuous* throughout the initiative. Too many change initiatives are undone because the participation of stakeholders is seen as something that can begin midway through the process. Participation should begin at the beginning, and opportunities to participate should continue throughout the initiative. When advisory panels evolve into advisory boards and a 5-year commitment is required, genuine participation is clear. Those responsible for future water reuse initiatives could consider forms of participation that evolve throughout the life of the initiative, from less formal to more formal governance arrangements, and from advisory to decision-making roles.

Third, "*facilitation*" is a key concept and the *skill to make participation effective*. Effective participation involves a "letting-go" of authority by traditional management bodies so that decision-making power can be shared among a broader stakeholder community. One way of grasping this transition is to think in terms of facilitation rather than management. Those with ultimate responsibility for water reuse initiatives will act as the promoters of a collaborative process, rather than as managers of a top-down process (Conklin, 2001).

Note that the term "facilitation" is used a bit differently here than in the section on consensus building. Here it is a general behavior and a leader's gentle guidance that does not impose norms or expectations. In the consensus-building section, it is the third-party facilitator role and the application of a flexible yet determined process to guide participants toward a decision.

5.4 Sharing Power and Maintaining Decision-Making Authority – Troy Hartley

It is possible to share power, while maintaining decision-making authority. It comes from clear *role definition* and *commitment to listen to others* in a public process. People are more likely to accept an "advisory" role or the knowledge that the outcome of a consensus process must proceed through a mandated public review process when there is clarity about their role in the process from the beginning.

At the same time, people want to know that they are taken seriously and that their views matter, just as the views of the professionals and experts in the field matter. Therefore, hear everyone out, capture all opinions, and respectfully discuss concerns. One way this commitment can be expressed is through government and organizational leaders who endorse and champion a decision-making process. It can also be conveyed by individuals who show respect for others' views, promote an honest approach to communicating information and views, and are not afraid to admit mistakes. Communicating a genuine commitment can be slightly complicated by statutory or administrative authority that grants one government agency the final decision-making authority. People seem to be able to detect when communication is not genuine, and it makes them suspicious and quick to reject or oppose information (Hartley, 2001).

5.5 Psychology and Information Sharing – Maureen Austin

When providing information to the public, *less is more*. Professionals tend to provide as much information as possible. The material presented can be overwhelming to an audience when the concepts are unfamiliar. This can cause confusion and antagonism, which disrupts meaningful interaction and communication. Present information on water reuse and conservation initiatives in small installments, with each installation building upon the previously presented information.

Furthermore, specialized resource professions often develop unique vocabulary common to the profession but uncommon in everyday public dialog. When presenting information about water reuse initiatives, limit the use of *jargon* and explain concepts in simple terms. This improves the interactions among professionals and the public.

People tend to pay more attention to exchanges that occur within familiar contexts. In addition, people do not want to appear foolish when learning new information or trying new behaviors (i.e., use of reclaimed water). Familiar formats and contexts help alleviate this concern. Sharing new, technical, or complex information with the public is enhanced when it is presented using *local and familiar communication channels* within social networks. This can be accomplished in several ways, including sharing information in neighborhood newsletters and publications, via neighborhood meeting venues, or using stories or testimonials shared between neighborhood residents, colleagues, and other social acquaintances.

Human thought and perception relies a great deal upon *images* to process information and learn new concepts and ideas. The popular expression "a picture is worth a thousand words" illustrates the tendency to rely on images in perception and understanding. Visual references to water shortages (e.g., photographs of low reservoir levels) and simple graphics (e.g., Figure 3-2, the water cycle) can help reinforce messages from professionals that stress the need to conserve water.

Keep these tips in mind when seeking information from the public. Many public hearings and meetings that are designed for the purpose of satisfying regulatory obligations and obtaining citizen input and tend to stifle rather than support information gathering. The structure of the meetings is often very formal, beginning with a presentation of the issues by professionals and followed by a public input period. If the issue is particularly complex (and most natural resource issues are), citizens may experience difficulty and frustration that can overshadow prime opportunities for sharing their insights and preferences. By making *use of community settings* where citizens regularly gather for neighborhood meetings and social functions, professionals may be less inclined toward formal presentation and can focus on dialog and communication with citizens about their concerns. The input gathered in community settings allows participants to feel informed and to be part of the process without being inconvenienced. This results in program guidelines and policies that are more closely aligned with public interests.

Seeking public input via surveys, focus groups, and dialog about resource management plans allows individuals to state their *preferences* without having to fully understand the complexities of the issue. Giving the public options to choose from and rate according to their preference provides professionals with public input, encourages meaningful public involvement, and discourages the information overload that often results when well-meaning professionals overwhelm the public with too much information (Austin, 2001).

5.6 Risk Communication and Motivating People to Engage – Mary McDaniel

Communication experts have been working on water and other natural resource and environmental issues for years and have addressed people's perceptions of risks and their willingness to engage with public agencies. Many suggest that public dialog often is dominated by a few true believers who hold extreme views at either end of the spectrum and is usually competitive and antagonistic. Consequently, getting people involved and keeping them involved through a protracted and often tedious bureaucratic process is challenging. While there are no silver bullets, following are some practical tips to encourage effective communication.

Table 5-2. Tips for Effective Communication

Do	Don't
♦ Involve the public early, before all decisions have been made.	♦ Make all the decisions ahead of time.
♦ Be flexible	♦ Be inflexible.
♦ Listen and take notes.	♦ Be condescending.
♦ Feed people.	♦ Skip the refreshments or neglect to feed people and then say, "People will think we are trying to buy them with a cookie."
♦ Share control by allowing questions and deviation from the agenda.	♦ Make people hold their questions until the end of a 1.5 hour presentation
♦ Show empathy.	♦ Be defensive.
♦ Understand the psychological significance of the issue to the public. For example, agencies often underestimate the symbolic significance water has for their constituents. For many, water represents innocence, and telling them where it comes from is a fall from grace.	♦ Be technical.
♦ Remember that this is a process, and keep people updated.	♦ Offer the euphemism "secondary effluent" in place of treated wastewater and hope the public won't make the connection.
♦ Follow up all requests for additional information.	♦ Ignore follow-up responsibilities.
♦ Respect the public.	♦ Forget to turn off cell phones.
♦ Be available. Give people contact information and encourage them to use it.	♦ Have presenters wear suits and ties and sit on stage behind a draped white table.
♦ Recognize the importance of "fairness" in substantive discussions, such as siting, as well as in process.	
♦ Take an interest. Talk to people. Allow a little vulnerability.	
♦ Encourage tours to take the mystery and the "yuck factor" out of the treatment process.	
♦ Involve the press. Present the human interest side of the issue. Provide graphics and back-up materials.	

Finally, conduct *community assessments* in addition to *customer surveys*. For example, don't ask only what your customer wants in a water product, but also ask what the community cares about. If opposition emerges later in an initiative it will be because something the community cares about is being threatened or ignored, not because something the customer wants is not being provided (McDaniel, 2001)

5.7 Conclusion

This chapter brought together a range of perspectives and supplied general tips and issues to consider when tailoring public outreach, education, and participation activities. After conducting a community assessment and committing to the framework principles outlined in Chapter 3.0, these tips and the public participation techniques in the following chapter provide the basis for planning, designing, implementing, and monitoring a public outreach, education, and participation program.

CHAPTER 6.0

PUTTING THE PRINCIPLES INTO ACTION: PUBLIC PARTICIPATION TECHNIQUES

While the best public outreach, education, and participation strategies, tools and techniques will be designed based upon a comprehensive community assessment, specific strategies, tools, and techniques do not need to be created from scratch. Many approaches have been used, tested, and refined over the years in related water and other natural resource management settings. This chapter provides a summary of those techniques.

However, while many public participation options exist, one consistent message emerges from the research that developed this framework guidance – *use all methods practicable*. It is not a matter of choosing between options, but rather employing all or as many as possible and tailoring them to the community and the framework principles.

6.1 Approaches to Public Participation

In selecting public outreach, education, and participation options, government agencies and water utilities need to give considerable thought to the question of success. What are their goals? The most problematic goals can be the unstated ones, such as implicit assumptions that "good process" will lead to "good outcomes." This often leads to an emphasis on participatory forms (e.g., advisory committees, public meetings, citizen juries), which, while important, may be less important than how the participatory forms are used (and for what purpose).

For example, advisory committees have generated ill will and failed to lead to solutions because the sponsoring agency was seeking ratification of proposals rather than "advice." However, in other situations advisory committees have been a means of collaborative problem solving. In short, an advisory committee is merely a tool. The tool can be used expertly or poorly for a range of functions from soliciting ideas to transmitting information. Similarly, although public meetings occasionally devolve into shouting matches, in many situations and if properly structured, meetings can be useful for generating ideas.

It can be problematic to rely on only one participatory "tool" for large or controversial participatory efforts. For example, over reliance on advisory committees may risk isolating representatives from their constituencies' interests and may result in decisions that the constituencies' cannot support.

Instead, a variety of participatory mechanisms may be needed for a variety of audiences. For example, a state agency that developed water quality regulations formed an advisory committee that met for a day a month. However, the agency also held informal meetings with a variety of stakeholder groups, developed a thought-provoking "survey" for stakeholders to consider tradeoffs, provided copy for newsletters, sent updates to mailing lists, etc. In this way, the agency was able to exchange ideas with a range of people – instead of relying on opinions of a few (Chess, 2001).

See the following tables from the International Association for Public Participation (2000), that summarize public participation techniques.

Table 6-1. The IAP2 Public Participation Spectrum

THE IAP2 PUBLIC PARTICIPATION SPECTRUM

Developed by the International Association for Public Participation

INCREASING LEVEL OF PUBLIC IMPACT →

Inform	Consult	Involve	Collaborate	Empower
Objective: To provide the public with balanced and objective information to assist them in understanding the problem, alternatives, and/or solution.	**Objective:** To obtain public feedback on analysis, alternatives and/or decisions.	**Objective:** To work directly with the public throughout the process to ensure that public issues and concerns are consistently understood and considered.	**Objective:** To partner with the public in each aspect of the decision including the development of alternatives and the identification of the preferred solution.	**Objective:** To place final decison–making in the hands of the public.
Promise to the Public: We will keep you informed.	**Promise to the Public:** We will keep you informed, listen to and acknowledge concerns, and provide feedback on how public input influenced the decision.	**Promise to the Public:** We will work with you to ensure that your concerns and issues are directly reflected in the alternatives developed and provide feedback on how public input influenced the decision.	**Promise to the Public:** We will look to you for direct advice and innovation in formulating solutions and incorporate your advice and recommendations into the decisions to the maximum extent possible.	**Promise to the Public:** We will implement what you decide.
Example Tools Fact sheets Web sites Open houses	**Example Tools** Public comment Focus groups Surveys Public meetings	**Example Tools** Workshops Deliberative polling	**Example Tools** Citizen Advisory Committees Consensus-building Participatory decision-making	**Example Tools** Citizen Juries Ballots Delegated decisions

visit www.iap2.org

Table 6-2. IAP2 Public Participation Toolbox: Passive Public Information Techniques

IAP2 Public Participation Toolbox

Tools and Techniques	Always Think It Through	What Can Go Right	What Can Go Wrong
PRINTED PUBLIC INFORMATION MATERIALS - Fact Sheets - Newsletters - Brochures - Issue Papers	• KISS! - Keep It Short and Simple • Make it visually interesting but avoid a slick sales look • Include a postage-paid comment form to encourage two-way communication and to expand mailing list • Be sure to explain public role and how public comments have affected project decisions • Q&A format works well	• Can reach large target audience • Allows for technical and legal reviews • Encourages written responses if comment form enclosed • Facilitates documentation of public involvement process	• Only as good as the mailing list/ distribution network • Limited capability to communicate complicated concepts • No guarantee materials will be read
INFORMATION REPOSITORIES Libraries, city halls, distribution centers, schools, and other public facilities make good locations for housing project-related information	• Make sure personnel at location know where materials are kept • Keep list of repository items • Track usage through a sign-in sheet	• Relevant information is accessible to the public without incurring the costs or complications of tracking multiple copies sent to different people • Can set up visible distribution centers for project information	• Information repositories are often not well used by the public
TECHNICAL REPORTS Technical documents reporting research or policy findings	• Reports are often more credible if prepared by independent groups	• Provides for thorough explanation of project decisions	• Can be more detailed than desired by many participants • May not be written in clear, accessible language
ADVERTISEMENTS Paid advertisements in newspapers and magazines	• Figure out the best days and best sections of the paper to reach intended audience • Avoid rarely read notice sections	• Potentially reaches broad public • May satisfy legal notification requirements	• Expensive, especially in urban areas • Allows for relatively limited amount of information
NEWSPAPER INSERTS A "fact sheet" within the local newspaper	• Design needs to get noticed in the pile of inserts • Try on a day that has few other inserts	• Provides community-wide distribution of information • Presented in the context of local paper, insert is more likely to be read and taken seriously • Provides opportunity to include public comment form	• Expensive, especially in urban areas

Table 6-2. IAP2 Public Participation Toolbox: Passive Public Information Techniques (cont.)

IAP2 Public Participation Toolbox

Tools and Techniques	Always Think It Through	What Can Go Right	What Can Go Wrong
FEATURE STORIES Focused stories on general project-related issues	• Anticipate visuals or schedule interesting events to help sell the story • Recognize that reporters are always looking for an angle	• Can heighten the perceived importance of the project • More likely to be read and taken seriously by the public	• No control over what information is presented or how
BILL STUFFER Information flyer included with monthly utility bill	• Design bill stuffers to be eyecatching to encourage readership	• Widespread distribution within service area • Economical use of existing mailings	• Limited information can be conveyed • Message may get confused as from the mailing entity
PRESS RELEASES	• Try to hand deliver press releases or kits to get a chance to discuss project • Foster a relationship with editorial boards and reporters	• Informs the media of project milestones • Press release language is often used directly in articles • Opportunity for technical and legal reviews	• Generally low media response rate • Frequent poor placement of press release within newspapers
NEWS CONFERENCES	• Make sure all speakers are trained in media relations	• Opportunity to reach all media in one setting	• Limited to news-worthy events
TELEVISION Television programming to present information and elicit audience response	• Cable options are expanding and can be inexpensive • Check out expanding video options on the internet	• Can be used in multiple geographic areas • Many people will take the time to watch rather than read	• High expense • Difficult to gauge impact on audience
WEB SITES World wide web sites which contain project information, announcments, and documents	• Keep it simple and easy to navigate • Use a logical site organization • Always keep site up to date	• Capable of reaching very large audiences with enormous amounts of information • Can be a very low cost way of distributing larger documents	• Many people still cannot access the web • Information overload and poor design can prevent people from finding what they need

Table 6-3. IAP2 Public Participation Toolbox: Active Public Information Techniques

IAP2 Public Participation Toolbox

Tools and Techniques	Always Think It Through	What Can Go Right	What Can Go Wrong
BRIEFINGS Use regular meetings of social and civic clubs and organizations to provide an opportunity to inform and educate. Normally these groups need speakers. Examples of target audiences: Rotary Club, Lions Clubs, Elks Clubs, Kiwanis, League of Women Voters. Also a good technique for elected officials.	• KISS - Keep it Short and Simple • Use "show and tell" techniques • Bring visuals	• Control of information/ presentation • Opportunity to reach a wide variety of individuals who may not have been attracted to another format • Opportunity to expand mailing list • Similar presentations can be used for different groups • Can build community good will	• Project stakeholders may not be in target audiences • Topic may be too technical to capture interest of audience
CENTRAL INFORMATION CONTACT Designated contacts are identified as official liaisons for the public and media	• If possible, list a person not a position • Best if contact person is local • Anticipate how phones will be answered • Make sure all recorded messages are kept up to date	• People don't get "the run around" when they call • Controls information flow and promotes information consistency • Conveys image of "accessibility"	• Designated contact must be committed to and prepared for prompt and accurate responses • May filter public message from technical staff and decision makers • May not serve to answer many of the toughest questions
INFORMATION HOT LINE Identify a separate line for public access to prerecorded project information or to reach project team members who can answer questions/ obtain input, also use emal and web sites	• Make sure contact has sufficient knowledge to answer most project-related questions • If possible, list a person not a position • Best if contact person is local • Use toll free number if not local	• People don't get "the run around" when they call • Controls information flowand promotes information consistency • Conveys image of "accessibility" • Easy to provide updates on project activities	• Designated contact must be committed to and prepared for prompt and accurate responses
TECHNICAL ASSISTANCE Providing access to technical expertise to individuals and organizations	• The technical resource must be perceived as credible by the audience • Work with your technical people to make sure they understand public issues	• Builds credibility and helps address public concerns about equity • Can be effective conflict resolution technique where facts are debated	• Availability of technical resources may be limited • Technical experts may not be prepared for working with the public

Table 6-3. IAP2 Public Participation Toolbox: Active Public Information Techniques (cont.)

IAP2 Public Participation Toolbox

Tools and Techniques	Always Think It Through	What Can Go Right	What Can Go Wrong
SIMULATION GAMES Exercises that simulate project decisions	• Test "game" before using • Be clear about how results will be used	• Can be designed to be an effective educational/ training technique, especially for local officials	• Requires substantial preparation and time for implementation • Can be expensive
INFORMATION CENTERS and FIELD OFFICES Offices established with prescribed hours to distribute information and respond to inquiries	• Provide adequate staff to accommodate group tours • Use brochures and videotapes to advertise and reach broader audience • Consider providing internet access station • Select an accessible and frequented location	• Provides opportunity for positive media coverage at groundbreaking and other significant events • Excellent opportunity to educate school children • Places information dissemination in a positive educational setting Information is easily accessible to the public • Provides an opportunity for more responsive ongoing communications focused on specific public involve-ment activities	• Relatively expensive, especially for project-specific use • Access is limited to those in vicinity of the center unless facility is mobile
EXPERT PANELS Public meeting designed in "Meet the Press" format. Media panel interviews experts from different perspectives.	• Provide opportunity for participation by general public following panel • Have a neutral moderator • Agree on ground rules in advance • Possibly encourage local organ-izations to sponsor rather than challenge	• Encourages education of the media • Presents opportunity for balanced discussion of key issues • Provides opportunity to dispel scientific misinformation	• Requires substantial preparation and organization • May enhance public concerns by increasing visibility of issues
FIELD TRIPS Provide tours for key stakeholders, elected officials, advisory group members and the media	• Know how many participants can be accommodated and make plans for overflow • Plan question/ answer session Consider providing refreshments • Demonstrations work better than presentations • Make sure everything is safe	• Opportunity to develop rapport with key stakeholders • Creates greater public knowledge of issues and processes	• Number of participants is limited by logistics • Potentially attractive to protestors

Table 6-3. IAP2 Public Participation Toolbox: Active Public Information Techniques (cont.)

IAP2 Public Participation Toolbox

Tools and Techniques	Always Think It Through	What Can Go Right	What Can Go Wrong
OPEN HOUSES An open house to allow the public to tour at their own pace. The facility should be set up with several stations, each addressing a separate issue. Resource people guide participants through the exhibits.	• Someone should explain format at the door • Ask participants to fill out a comment sheet • Be prepared for a crowd all at once - develop a meeting contingency plan • Set up stations so that several people (6-10) can view at once	• Fosters small group or one-on-one communications • Ability to draw on other team members to answer difficult questions • Meets information and interaction needs of many members of the public who are not served by typical public meetings • Builds credibility	• Difficult to document public input • Protestors may use the opportunity to disrupt event • Usually more staff intensive than a meeting • May not provide the opportunity to be heard that some public will expect
COMMUNITY FAIRS Central event with multiple activities to provide project information and raise awareness	• All issues, large and small must be considered • Make sure adequate resources and staff are available	• Focuses public attention on one element • Conducive to media coverage • Allows for different levels of information sharing	• Public must be motivated to attend • Usually expensive to do it well • Can damage reputation if not done well

Table 6-4. IAP2 Public Participation Toolbox: Small Group Public Input Techniques

IAP2 Public Participation Toolbox

Tools and Techniques	Always Think It Through	What Can Go Right	What Can Go Wrong
INTERVIEWS One-to-one meetings with stakeholders to gain information on public concerns and perspectives for developing or refining public involvement and consensus building programs	• Where feasible, interviews should be conducted in-person, particularly when considering candidates for citizens committees • Take advantage of opportunity for public to input in how they participate	• Provides opportunity to get understanding of public concerns and issues • Provides opportunity to learn how to best communicate with public • Can be used to evaluate potential citizen committee members	• Scheduling multiple interviews can be time consuming • Interviewers must engender trust or risk negative response to format
IN-PERSON SURVEYS One-on-one "focus groups" with standardized questionnaire or methodology such as "stated preference"	• Make sure intended use of result is clear before technique is designed	• Provides traceable data • Reaches broad, representative public	• Expensive • Focus Groups may have a marketing/ public relations image
COFFEE KLATCHES Small meetings within neighborhood usually at a person's home	• Make sure staff is very polite and appreciative	• Relaxed setting is conducive to effective dialogue • Maximizes two-way communication	• Requires a lot of labor to reach many people
SMALL FORMAT MEETINGS Small meetings at existing groups or in conjunction with another event	• Understand who the likely audience is to be • Make opportunities for one-one-one meetings	• Opportunity to get on the agenda • Provides opportunity for in-depth information exchange in non-threatening forum	• May be too selective and can leave out important groups

Table 6-5. IAP2 Public Participation Toolbox: Large Group Public Input Techniques

IAP2 Public Participation Toolbox

Tools and Techniques	Always Think It Through	What Can Go Right	What Can Go Wrong
RESPONSE SHEETS Mail-In-forms often included in fact sheets and other project mailings to gain information on public concerns and preferences	• Use prepaid postage • Include a section to add name to the mailing list • Document results as part of public involvement record	• Provides input from those who would be unlikely to attend meetings • Provides a mechanism for expanding mailing list	• Does not generate statistically valid results • Only as good as the mailing list • Results can be easily skewed
MAILED SURVEYS & QUESTIONNAIRES Inquiries mailed randomly to sample population to gain specific information for statistical validation	• Make sure you need statistically valid results before making investment • Survey/ questionnaire should be professionally developed and administered to avoid bias • Most suitable for general attitudinal surveys	• Provides input from individuals who would be unlikely to attend meetings • Provides input from cross-section of public not just activists • Statistically tested results are more persuasive with political bodies and the general public	• Response rate is generally low • For statistically valid results, can be labor intensive and expensive • Level of detail may be limited • May be preceived as a public relations tool
TELEPHONE SURVEYS/ POLLS Random sampling of population by telephone to gain specific information for statistical validation	• Make sure you need statistically valid results before making investment • Survey/ Questionnaire should be professionally developed and administered to avoid bias • Most suitable for general attitudinal surveys	• Provides input from individuals who would be unlikely to attend meetings • Provides input from cross-section of public, not just those on mailing list • Higher response rate than with mail-in surveys	• More expensive and labor intensive than mailed surveys • Bias is easily charged if questions not carefully constructed
INTERNET SURVEYS/ POLLS Web-based response polls	• Be precise in how you set up site, chat rooms or discussion places can generate more input than you can look at	• Provides input from individuals who would be unlikely to attend meetings • Provides input from cross-section of public, not just those on mailing list • Higher response rate than other communication forms	• Generally not statistically valid results • Can be very labor intensive to look at all of the responses • Cannot control geographic reach of poll • Results can be easily skewed
COMPUTER-BASED PARTICIPATION Surveys conducted via computer network	• Appropriate for attitudinal research	• Provides instant analyses of results • Can be used in multiple areas • Novelty of technique improves rate of response	• High expense • Detail of inquiry is limited
PUBLIC HEARINGS Formal meetings with scheduled presentations offered	• Avoid if possible, otherwise try to use informal meetings immediate before	• Provides opportunity for public to speak without rebuttal • meets legal requirements • puts comments on record	• Does not foster dialogue • Creates us vs. them feeling • Many dislike public speaking

Table 6-6. IAP2 Public Participation Toolbox: Small Group Problem-Solving Techniques

IAP2 Public Participation Toolbox

Tools and Techniques	Always Think It Through	What Can Go Right	What Can Go Wrong
DESIGN CHARRETTES Intensive session where participants re-design project features	• Best used to foster creative ideas • Be clear about how results will be used	• Promotes joint problem solving and creative thinking • Effective for creating partnerships and positive working relationships with public	• Participants may not be seen as representative by larger public • May not have lasting effect if used as a one-shot technique
COMMUNITY FACILITATORS Use qualified individuals in local community organizations to conduct project outreach	• Define roles, responsibilities and limitations up front • Select and train facilitators carefully	• Promotes community-based involvement • Capitalizes on existing networks • Enhances project credibility	• Can be difficult to control information flow • Can build false expectations
MEDIATION / NEGOTIATION The process of resolving disputes through compromise	• Should be used typically as a last resort to solve specific problems with well-defined stakeholders groups	• Promotes accountability on both sides • Focuses on specific issues	• Difficulty of defining who the parties are and whom they represent • Time and labor intensive
CONSENSUS BUILDING TECHNIQUES Techniques for building consensus on project decisions such as criteria and alternative selection. Often used with advisory committees. Techniques include Delphi, nominal group process and public value assessment and many others.	• Use simplified methodology • Allow adequate time to reach consensus • Consider one of the computerized systems that are available • Define levels of consensus, i.e. a group does not have to agree entirely upon a decision but rather agree enough so the discussion can move forward • Make sure decision maker is committed to consensus	• Encourages compromise among different interests • Provides structured and trackable decision making • Focuses on solving problems with mutually satisfactory solutions • Can help avoid later conflicts	• Not appropriate for groups with no interest in compromise • Consensus may not be reached
FOCUS GROUPS Message testing forum with randomly selected members of target audience. Can also be used to obtain input on planning decisions	• Conduct at least two sessions for a given target • Use a skilled focus group facilitator to conduct the session	• Provides opportunity to test key messages prior to implementing program • Works best for select target audience	• Relatively expensive if conducted in focus group testing facility

Table 6-6. IAP2 Public Participation Toolbox: Small Group Problem-Solving Techniques (cont.)

IAP2 Public Participation Toolbox

Tools and Techniques	Always Think It Through	What Can Go Right	What Can Go Wrong
ADVISORY COMMITTEES A group of representative stakeholders assembled to provide public input to the planning process	• Define roles and responsibilities up front • Be forthcoming with information • Use a consistently credible process • Interview potential committee members in person before selection • Use third party facilitation • Make sure members communicate with their constituencies	• Provides for detailed analyses for project issues • Participants gain understanding of other perspectives, leading toward compromise	• General public may not embrace committee's recommendations • Members may not achieve consensus • Sponsor must accept need for give-and-take • Time and labor intensive
TASK FORCES A group of experts or representative stakeholders formed to develop a specific product or policy recommendation	• Obtain strong leadership in advance • Make sure membership has credibility with the public • Make sure members represent diverse perspectives and will be independent	• Findings of a task force of independent or diverse interests will have greater credibility • Provides constructive opportunity for compromise	• Task force may not come to consensus or results may be too general to be meaningful • Time and labor intensive
PANELS A group assembled to debate or provide input on specific issues	• Most appropriate to show different views to public • Panelists must be credible with public	• Provides opportunity to dispel misinformation • Can build credibility if all sides are represented • May create wanted media attention	• May create unwanted media attention • Can polarize issues if not conceived and moderated well
CITIZEN JURIES Small group of ordinary citizens empanelled to learn about an issue, cross examine witnesses, make a recommendation. Always non-binding with no legal standing	• Requires skilled moderator • Commissioning body must follow recommendations or explain why • Be clear about how results will be used	• Great opportunity to develop deep understanding of an issue • Public can identify with the "ordinary" citizens • Pinpoint fatal flaws or gauge public reaction	• Resource intensive
ROLE-PLAYING Participants act out characters in predefined situation followed by evaluation of the interaction	• Choose roles carefully. Ensure that all interests are represented. • People may need encouragement to play a role fully	• Allow people to take risk-free positions and view situation from other perspectives • Participants gain clearer understanding of issues	• People may not be able to actually achieve goal of seeing another's perspective

Table 6-7. IAP2 Public Participation Toolbox: Large Group Problem-Solving Techniques

IAP2 Public Participation Toolbox

Tools and Techniques	Always Think It Through	What Can Go Right	What Can Go Wrong
ELECTRONIC DEMOCRACY Internet, Websites, Televoting, On-line Dialogue, On-line Delivery of Government Services	• Carefully plan how information will be presented and how feedback will be used	• Facilitates interactive communication • Convenient	• Not accessible to everyone • Opportunity for manipulation/ misinformation/ incivility
SAMOAN CIRCLE Leaderless meeting that stimulates active participation	• Set room up with center table surrounded by concentric circles • Need microphones • Requires several people to record discussion	• Can be used with 10 to 500 people • Works best with controversial issues	• Dialogue can stall or become monopolized
OPEN SPACE TECHNOLOGY Participants offer topics and others participate according to interest	• Important to have a powerful theme or vision statement to generate topics • Need flexible facilities to accommodate numerous groups of different sizes • Groundrules and procedures must be carefully explained for success	• Provides structure for giving people opportunity and responsibility to create valuable product or experience • Includes immediate summary of discussion	• Most important issues could get lost in the shuffle • Can be difficult to get accurate reporting of results
WORKSHOPS An informal public meeting that may include a presentations and exhibits but ends with interactive working groups	• Know how you plan to use public input before the workshop • Conduct training in advance with small group facilitators. Each should receive a list of instructions, especially where procedures involve weighting/ ranking of factors or criteria	• Excellent for discussions on criteria or analysis of alternatives • Fosters small group or one-to-one communication • Ability to draw on other team members to answer difficult questions • Builds credibility • Maximizes feedback obtained from participants • Fosters public ownership in solving the problem	• Hostile participants may resist what they perceive to be the "divide and conquer" strategy of breaking into small groups • Several small-group facilitators are necessary
FUTURE SEARCH CONFERENCE Focuses on the future of an organization, a network of people, or community	• Hire a facilitator experienced in this technique	• Can involve hundreds of people simultaneously in major organizational change decisions • Individuals are experts • Can lead to substantial changes across entire organization	• Logistically challenging • May be difficult to gain complete commitment from all stakeholders • 2 – 3 day meeting
DELIBERATIVE POLLING Measures informed opinion on an issue	• Do not expect or encourage participants to develop a shared view • Hire a facilitator experienced in this technique	• Can tell decision-makers what the public would think if they had more time and information • Exposure to different backgrounds, arguments, and views	• Resource intensive • Often held in conjunction with television companies • 2 – 3 day meeting

6.2 Marketing Reclaimed Water

At the August 2000 water reuse symposium, many participants were interested in finding better methods for marketing reclaimed water and increasing demand for a valued product through designing and delivering more effective messages. At the same time, others expressed concern that marketing is not inherently genuine, and may represent an underhanded attempt to manipulate the public.

While misperceptions abound about what marketing is and can do on both sides of the argument, marketing strategies play an important role but must be managed with care, like all public outreach, education, and participation activities. Marketing strategists who attended the symposium described marketing as:

1. Identifying what the public values.
2. Determining how a product or item addresses those values.
3. Crafting a communication strategy that emphasizes the values and product's relationship to the public value.

Many marketing professionals believe marketing strategies are consistent with participatory mechanisms. If marketing is viewed in whole, rather than simply the third phase of crafting "slick" messages, then it compliments participatory methods. Using highly participatory methods to identify and understand public values and to assess how reclaimed water might address those values is a prerequisite to crafting effective communication strategies. Jumping immediately into crafting a message could make the public perceive utilities as disingenuous and false-hearted. The public can detect when someone is not genuine, and a utility can do tremendous damage to its credibility and the public's confidence by using an uninformed communication strategy, that is, by prematurely crafting messages before assessing the public's values.

Marketing strategies play an important role in an overall program of communicating with the public about water reuse. But, know the risks and understand the misperceptions that are associated with the word "marketing." Anything that suggests manipulation of the public or not taking their interests, concerns, or value seriously is very risky and can damage credibility and the public's confidence in a utility. Therefore, resist the temptation to jump immediately to designing public messages and communication strategies before fully understanding the community's values and the relationship that reclaimed water has with those values. After conducting a comprehensive community assessment and implementing ongoing feedback measures, then using marketing approaches to design communication strategies can be very helpful.

6.3 Advisory Committees – Tips from Former Tucson, Arizona, City Council Member, Margot Garcia

Blue ribbon panels or advisory committees are often established to discuss controversial and complex issues with the best technical advice available, both from the local government as well as from impartial professionals. Panels and committees write recommendations. This section on advisory committees features comments from a former city council member who was recalled, along with other city council members, after a controversial water decision in Tucson, Arizona.

Advisory committees can be useful forums for getting opinion leaders in the community to hear the same technical information that agency staff and elected officials hear and to work through issues to a conclusion. Constructing such a committee has to be done carefully and risks exists for the members and the local government. The issues that have to be handled carefully are:

- Who will be asked to serve?
- Who will chair the committee?
- What will be the length of appointment?
- Who will set the agenda?
- How will the resultant report be issued?

6.3.1 Representation

The committee can be assembled in a number of ways. For example, each council member/supervisor may be asked to appoint one member (that builds some political buy-in into the process), with the city or county manager appointing three or four people. The city or county manager has to be sure to balance the political appointments with people who are either the critics (if the elected officials did not appoint them), leaders of important interest groups (e.g., chamber of commerce, large local industry, large environmental group, neighborhood or civic association council, cultural or ethnic groups, and "movers and shakers"), and impartial professionals (e.g., university professors or members of professional organizations). The group needs to have a balance of men and women, racial and ethnic minorities, and geographic areas of the city or county. Creating a committee without critics or without minorities will only make the committee's recommendations subject to criticism based on a lack of fairness and opportunity to be heard. Assembling the group can be difficult. All the people invited to serve need to be willing to put in the time and effort and recognize that this will be a working group with frequent meetings.

6.3.2 Leadership

The chair should be appointed by the mayor or chair of board of supervisors. This person should be someone who is committed to making this effort work, who has good leadership skills, and who is respected throughout the community. One reason for appointing a chair, as opposed to electing the chair from within the group, would be to avoid early battles. Elections could create the perception that cliques already are forming in within the advisory committee. This could erode good will.

Citizens who participate risk of being co-opted by the prestige of serving on the committee, the camaraderie of the group, and the sense that they are making the decision. The committee always has to remember that it is advisory to the elected officials. However, the members who are appointed by an elected official have a door open to the people who appointed them and should take advantage of it to keep the elected official informed about what is happening and to lobby.

6.3.3 Length of Term

For most complex issues, it will take a year to write a report, even if meetings are held every month. To take only 6 months will require meeting twice a month. Therefore, people should be appointed for the duration of the task. If the group is working together successfully and the elected officials feel it has something to offer over the long term, then the members can be asked if they wish to continue for 1 or 2 years, and staggered terms should be set up. People should serve no longer than 4 years. Serving longer than that, the group gets stale and becomes a rubber stamp. It needs new blood to ask the question, "Why do you do that?" and someone to challenge the assumptions that underlie the decision-making process.

6.3.4 Agenda Setting

The staff and chair should set the agenda for each meeting. The chair will know what the members think is important to discuss. The staff members will know the resources available for education and will also bring forward topics they think are important. Just having the staff set the agenda makes the committee feel like it is just reacting and has no control over what it wants to do. The chair also needs to bring to the agenda topics that the committee members have requested for inclusion.

The role of outside professionals (i.e., non-governmental staff) is important in giving the committee "unbiased" information and using the best science available. The staff has to be prepared to listen and handle opinions that differ from their own. They can debate with the outside professional in front of the committee. That gives the committee an opportunity to listen to both sides and establish their own opinions. Often, the problem is the underlying assumptions, and that is important for both committee and staff to understand. The staff has to be careful not to undermine or disparage alternative opinions. The chair's handling of the meeting is very important when the staff interacts with outside professionals.

6.3.5 Final Report

The purpose of the report is to advise the elected officials on a policy issue. The committee has to remember that it is only issuing advice. The elected officials can adopt it in whole, in part, or not at all.

Who writes the report is always a big issue because the committee members want control over what is written but often do not have the time to actually do the writing. However, the committee members may want to write specific sections. So, the committee as a whole should set up the outline for the report and determine who will write each section. The staff should fill in the rest. The report needs to be lightly edited (i.e., errors need to be corrected) by staff. However, the staff should not to change meaning and intent. The document needs to be read by the committee and agreed to by vote. If the vote is not unanimous, the negative votes should be allowed to write dissenting opinions (i.e., why they could not support the document), and those opinions should be made part of the document.

A press release should be sent out when the document is finished. Copies of the document should be placed in libraries and online and be made available to everyone upon request. The report should be seen as a document for discussion. The elected officials have to be prepared to get results they do not like.

In closing, advisory committees can be very constructive and effective tools in the public outreach, education, and participation arsenal. They benefit from close management attention and can be kept further informed through ongoing diagnostic activities that monitor the community (Garcia, 2001).

6.4 Conclusion

The best public outreach, education, and participation strategies, tools and techniques will be designed based upon a comprehensive community assessment that tailors the application of the principles to a specific context. At the same time, it is likely that the principles and the context will mandate the use of as many public outreach, education, and participation approaches as feasible, prioritized for the community.

CHAPTER 7.0

Conclusion

People will forgive you for your mistakes; but they won't forgive you for not telling them about them.
—Risk Communication Expert, 2001

So called "people problems" cannot be solved with a newly engineered filter, a bioengineered microbe, or even a rational analysis of the scientific knowledge by a panel of blue ribbon experts. Rather, the people side of water reuse must be addressed through hand-shakes, coffee and donuts, and frequent, varied types of communication. The people side is about managing all forms of information, from data through to the expression of respect, in a genuine interpersonal interaction. Over time, this builds a public perception of fairness and trust in the water utility. Once public confidence and trust is on the mend, tackling challenging, contentious public decisions can take place in a more constructive decision-making environment. It will continue to be messy at times, but progress can be made.

The framework presents a strategy for building public confidence and trust through a principle-based approach to public outreach, education, and participation. It requires two steps:

1. The application of diagnostic tools to generate a comprehensive community assessment, and the implementation of an ongoing community assessment monitoring program.
2. The tailoring of specific public outreach, education, and participation activities based upon the outcome of the community assessments *and* consistent with the five underlying principles.

The diagnostic step calls for an analysis of contextual factors specific to the community, along a scale from societal to individual levels. Specifically, the contextual factors include:

- Societal and institutional context
- Group profile and decision context
- Individual characteristics

Analysis of data collected on these contextual factors aims to identify important patterns and trends. It is the analysis of the contextual data that enables a utility to tailor its products and activities.

Second, once an understanding of the context exists and a monitoring system is in place to help detect changes in the context, apply the five principles to the context to determine specific public outreach, education, marketing, and participation strategies, tools, and techniques. This exercise can be applied at several levels of management, from selecting overall communication strategies to designing and implementing individual participation activities (e.g., creating citizen's advisory committees or a marketing message). The framework principles reiterated below.

7.1 Principle 1: Manage Information for All

Diverse information should be managed to promote learning, communication, and mutual understanding among all the stakeholders, given that different people learn and communicate differently.

- Ensure equal access to information.
- Provide multiple methods of information presentation and communication.
- Support and adhere to individuals' information processes and tendencies.

7.2 Principle 2: Maintain Individual Motivation and Demonstrate Organizational Commitment

Individuals should find more good reasons (e.g., benefits for self and community) than bad reasons (e.g., real or perceived risks) to engage in public dialogue. Organizations should demonstrate genuine commitment to engaging the public.

- Ensure and nurture multiple motives for the public to engage, including individual and community benefits.

Demonstrate genuine commitment throughout the organization to hearing the public and its concerns.

7.3 Principle 3: Promote Communication and Public Dialog

Communication and the broader, cumulative public dialog should take multiple forms or venues and be on-going at all stages of a decision-making process.

- Expect to have to say the message multiple times (and in multiple ways) before it is truly understood by others.
- Expect to have to listen to others' messages multiple times and in multiple ways before truly understanding them.
- The quality of the public dialogue is an indicator of the utility's relationship with the public and the public's confidence in the utility.

7.4 Principle 4: Ensure Fair and Sound Decision-Making and Decisions

Participants should perceive the decision-making process and the outcome as fair and sound.

- Fairness includes process or procedural dimensions and outcome or distributive elements.
- Sound decisions are reasoned, well thought-out, and based upon accepted knowledge.

7.5 Principle 5: Build and Maintain Trust

The cumulative attention to the previous principles contributes to the establishment and maintenance of public confidence and trust.

- Build trust and credibility reserves with the public and tap into them when necessary.

The *Framework* truly is guidance, *not* a checklist of "to-dos" to establish public confidence. Quite the opposite, research suggests that a one-size-fits-all approach cannot work because the most appropriate ways to apply the core principles can vary from case to case. Rather, what is critical is the utility's commitment to principles and capabilities to tailor them based upon the context of the community, the capabilities of the participants, and the decision at hand. Context and capabilities change over time. Consequently, the tools that are most successful also must change over time.

7.6 Additional Resources

To delve deeper into multiple dimensions of the societal context briefly presented in this framework, examine the literature review (Appendix A). In addition, many resources about public outreach, education, and participation are available to water professionals; many are available at the local level. Following are a few resources that were identified during the development of the framework. These references include further tips and some tools and techniques for conducting public outreach and education and for increasing participation.

7.6.1 Public Participation Manuals and Other Guidance Documents

- Adler, P.S., R.C. Barrett, M.C. Bean, J.E. Birkhoff, C.P. Ozawa, and E.B. Rubin. *Managing Scientific and Technical Information in Environmental Cases: Principles and Practices for Mediators and Facilitators*, June 2000, Washington DC: RESOLVE, Inc., U.S. Institute for Environmental Conflict Resolution, and the Western Justice Center Foundation. www.resolv.org/resources/pub/default.htm
- American Water Works Association Research Foundation (AWWARF). 1995. *Public Involvement Strategies: A Managers Handbook*. Washington DC: AWWARF.
- American Water Works Association Research Foundation (AWWARF) 2001. *Public Involvement: Making It Work*. Washington DC: AWWARF.

- Bleiker, A., and H. Bleiker. *Public participation handbook for officials and other professionals serving the public*, 9th ed. Monterey, CA: Institute for Participatory Management and Planning.
- Bryan, T.A. 1997. *Effective Stakeholder Involvement Strategies for Environmental Technology Projects: Principles and Practical Applications*, January. Denver: Colorado Center for Environmental Management.
- Chess, C., B.J. Hance, and P.M. Sandman. 1988. *Improving Dialogue with Communities: A Short Guide for Government Risk Communication*, January. New Brunswick, NJ: Environmental Communication Center, Rutgers University.
- Environmental Resources Management. 1995. *Manual on public participation.* Report prepared for the European Bank for Reconstruction and Development, London UK. Environmental Resources Management.
- U.S. Environmental Protection Agency. 1998. *Community Advisory Group Toolkit*, EPA 540-R-97-037, September. (http://www.epa.gov/superfund/tools/cag/cagtlktc.pdf) Washington, DC: Office of Solid Waste and Emergency Response.

Also see Appendix A for a literature review of studies and additional publications.

7.6.2 Web Sites

- Colearning and Community Involvement:
  - EnviroTools – educational resources. www.envirotools.org/
- Community Profiling:
  - Evaluation Resources Toolbox for Stakeholder Involvement. www.epa.gov/evaluate/toolbox/index.htm
- Dispute Resolution and Consensus-Building:
  - RESOLVE, Inc. www.resolv.org
  - U.S. Institute for Environmental Conflict Resolution. www.ecr.gov
  - The Consensus Building Institute, Inc. www.cbi-web.org
- Environmental Decision Making:
  - National Center for Environmental Decision-Making Research. www.ncedr.org
- Environmental Outreach and Education:
  - Environmental Education on the Internet. eelink.net
  - North American Association for Environmental Education. www.naaee.org/index.php
- Public Participation:
  - International Association for Public Participation. www.iap2.org
- Risk Communication:
  - Society for Risk Analysis. www.sra.org/

7.7 The Next Steps

This framework lays the groundwork for the principles essential to building the necessary public confidence, trust, and support for a wastewater utility's decision making. The framework describes a set of community assessment questions that inform a utility of the community-specific context in which a water reuse decision is taking place. With this information, it will be possible to tailor public outreach, education, marketing, and participation activities to meet the needs of a specific community. Only through the tailored approach can the framework principles be achieved and in turn, water reuse initiatives have the best chance of getting successfully implemented because of the establishment of public trust and confidence.

However, application of the framework strategy should be thoroughly tested in the water reuse context. Results should be adjusted accordingly over time to make the framework of greatest use to the water utility professional over the long term. WERF and utility members are in the best position to continue the application, testing, and refinement of the framework. At the same time, the National Research Council (NRC) is undertaking a "Study of Public Participation in Environmental Assessment and Decision Making" aimed at better understanding the factors that are related to "success" in participatory methods of environmental decision making. The efforts of the water utility research on framework principles should be coordinated with the NRC activities.

One strategy for testing would include:

- Developing and testing a set of diagnostic measures to produce a comprehensive community assessment and testing them in several real-world settings
- Creating detailed how-to information and guidance for applying the measures (data collection and analysis) and tailoring existing public outreach, education, marketing, and participation activities
- Designing and maintaining a Web site with additional background, guidance, and reference materials and a Web-based case database to continually collect contextual data on water reuse cases across the country in order to verify and refine the framework principles.

Utilities can achieve an unparalleled level of public confidence and be viewed by the public as an essential component of sustainable water resource management in America. However to reach this level of public confidence, a utility must be committed to building and maintaining a relationship with the public. While there are no quick fixes, the framework is a resource to help build public confidence and trust.

APPENDIX A

PUBLIC PERCEPTION & PARTICIPATION IN WATER REUSE:

LITERATURE SUMMARY

JULY, 2001

Troy W. Hartley, Ph.D.

TABLE OF CONTENTS

Table of Contents i

Introduction A-1
What is Water Reuse & What Do People Think of It? A-2
What Does Social Science Say About Perception & Participation? A-4
Literature Profile
Overarching Themes
What Does All This Tell Us? A-8
Next Steps

Attachment 1: What Is Behind These Five Themes? The Devil is in the Details A-10
Category 1: Individual Level Factors
Attitudes
Commitment & Motivation
Communicative Skills
Demographics
Interests, Concerns & Values
Knowledge & Awareness
Learning Abilities
Personal Behavior
Power
Psychological Processes
Risk Perception
Trust

Category 2: Decision-Making & Situational Context Level Factors
Complex Information
Dialog Quality
Empowerment
Fairness
Leadership
Process Features
Representation

Category 3: Institutional & Societal Context Factors
Institutional Authority
Organizational & Professional Cultures
Organizational Commitment
Social Capital

Attachment 2: Reference List

I. Introduction

Water is a limited resource in a rapidly expanding global population. Reclaiming water after it is treated in a modern wastewater treatment plant has been promoted as an element of sustainable water resource management strategies. Water reuse for non-potable (i.e., non-drinking) purposes is widely accepted, with communities using reclaimed water for a range of activities, including irrigate golf courses and landscaping, cooling waters for industry, and canals, ponds, and other aesthetic water features. However, in a growing number of cases, communities have considered using highly treated municipal wastewater to augment drinking water supplies. While engineering and science have achieved great success in addressing human health and environmental issues with reclaimed water, many water reuse initiatives have been terminated due to lack of public support. The use of reclaimed water as a resource is dependent upon public support. Why that support does not exist and how to best engage in a public discussion about water reuse involves gaining a better understanding of public perception and the development of effective mechanisms for public participation. RESOLVE, Inc, is considering these questions, with support from the Water Environment Research Foundation (WERF), in cooperation with the National Water Research Institute, American Water Works Association Research Foundation, and the WateReuse Foundation.

The purpose of this literature review summary is to examine a variety of perspectives on perception and public participation in complex environmental issues, and in particular, issues of water reuse. The literature review summary is the first step in a three-staged process of distilling the key factors that influence public perception and participation in a water reuse decision. For our purposes, a factor is any characteristic, phenomenon, or fact that influences in some way the perception held by individuals or the nature of the public participation experience. The literature summary, along with case studies and short white papers from experts in perception and participation, will serve as background material to stimulate discussion in a symposium held in Washington, DC in the summer of 2001. The symposium will bring together academic experts with water reuse professionals and stakeholders for a facilitated discussion. Together, researchers and practitioners will identify the most critical factors and develop strategies for addressing those factors. Finally, a *Framework* document will be written that bridges disciplinary and stakeholder perspectives and supplies practical, effective guidance and tools for water resource professionals. The *Framework* will also identify gaps in our current social science understanding of the factors underlying perception and public participation.

The summary begins with a brief introduction to water reuse, including the literature on how people perceive water reuse. It then shifts to the broader social science literature on perception and public participation, characterizing that literature and providing an overview of the major themes emerging from the diverse social science perspectives. The concluding section discusses the results of the small number of larger, comprehensive empirical studies in the field to date.

The first attachment (Attachment 1) breaks down the over-arching themes into more specific factors, discussing in greater detail the nature of twenty-three factors. The factors are organized into individual, decision context, and institutional and societal context categories. Attachment 2 is the complete reference list reflected in this literature review summary.

II. What is Water Reuse & What Do People Think of It?

Water reuse, reclaimed water, recycled water, repurified water are all terms for the beneficial use of water after it has been treated in a state-of-the-art, multi-stage wastewater treatment process. Worldwide, water is reused for industrial purposes, landscaping, agricultural irrigation, and indirect and direct potable uses. In the U.S., reclaimed water has been used predominantly for non-potable (i.e., non-drinking) purposes, with only a few indirect potable examples. As early as the 1930s, reclaimed water was discharged on to the ground and allowed to recharge the groundwater aquifer that serves as a drinking water supply in California – i.e., indirect potable reuse. In the late 1950s and early 1960s, additional indirect potable reuse projects in California and Nevada were initiated, as was a temporary, drought-driven direct potable reuse in Kansas – i.e., reclaimed water being directly routed to a drinking water treatment plant. By the mid-1970s, Virginia had an indirect potable reuse project. The 1980s and 1990s were a period of rapid expansion of water reuse, as communities in Colorado, California, Texas, Florida, and Arizona considered indirect potable uses. The 2000s promise additional initiatives, particularly as population and economic growth pressures further tighten water supplies. However, at the same time public has become increasingly uncomfortable and less accepting of indirect potable projects.[1]

The public supports the general concept of reusing reclaimed water, and are highly supportive of non-potable reuse. People favor reuse that promotes water conservation, provides environmental protection benefits, protects human health, and cost effectively treats and distributes a valuable and limited resource. However, as the water reuse option becomes more concrete, with specific proposed projects in communities, and the proposed use increases the likelihood of direct human contact, attitudes change – the public becomes less supportive. Attitude surveys from the 1970s and 1980s identified five factors contributing to the public's acceptance of water reuse initiatives. As a sixth factor, recent surveys in Orange County, California, where a water reuse initiative is currently underway, revealed that the public was concerned by the fact that the reclaimed water was originally wastewater.[2] Surveys in the UK confirmed this "source" factor, with people more willing to use recycled water from their own wastewater than from second parties or a common public source.[3] Thus, sources may be an additional acceptance factor:

- Degree of human contact
- Protection of public health
- Protection of the environment
- Promotion of water conservation
- Cost of treatment and distribution technologies and systems[4]
- Perception of waste water as source of reclaimed water

Case studies from the 1970s and 1980s identified additional awareness and trust factors influencing whether the public would accept the use of reclaimed water:

- Awareness of water supply problems
- Perception of role of reclaimed water in overall water supply allocation scheme
- Perception of the quality of reclaimed water
- Confidence in local management of public utilities and technologies[5]

In general, trust and confidence in public agencies and officials is in decline in America.[6] Likewise, belief that even the best technologies can remove all impurities and germs from waste water is in decline, according to a California survey.[7] Furthermore while the public has reported trusting university-based scientists and the medical community on technical and health issues related to water reuse, a recent UK survey showed people trusting their own personal impressions of water quality (often based upon the water's cloudiness or turbidity) more than these experts.[8]

In the 1990s, a number of high profile indirect potable water reuse projects encountered stiff public opposition. The public effectively organized themselves politically and prevented projects from being implemented in these cases. Terms like "Toilet to Tap" and "Sewage Beverage" framed the political discussion. Within this context, recent surveys have confirmed and expanded some of the underlying factors that contribute to these perceptions and concerns about water reuse. Demographic indicators have been identified:

Opponents tend to be:

- Seniors
- Long-time residents
- Lower education attainment
- Lower income

Proponents tend to be:

- Younger
- Greater education attainment
- Higher income[9]

There have been studies that claim men are more supportive than women in water reuse; however, other studies have concluded the opposite.[10] Overall, demographic factors may be weak indicators and unreliable if not confirmed by other measurement instruments. In addition, while information sharing, educational activities, and opportunities for reflection upon the concept of water reuse do increase support, it has also been shown to intensify the extremes. In other words, those that are opposed to water reuse are more strongly opposed and those that support it are more strongly supportive after becoming more knowledgeable and aware.[11]

In spite of serious opposition, the public has expressed an interest in being meaningfully involved in water reuse decision-making, and finding ways to ensure an independent and secure water supply for their communities.[12] The public is, in a very general sense, aware that there are water supply problems in many parts of the country and a few believe that some form of potable reuse is inevitable, given growth and water supply constraints.[13]

Within this context, the water reuse community has asked for greater clarification as to *why* the public holds these perceptions of reuse and *what* can be done to discuss these perceptions through public participation efforts. The first step in a comprehensive effort to examine the crucial elements of ***public perception*** and ***participation*** that underlie the public's opposition and behavior includes the broader review of the social science literature reported here. Subsequent steps include case studies, a symposium in the summer 2001, and a final framework report.

III. What Does Social Science Say About Perception & Participation?

Public perception and participation have been widely discussed in a broad array of social science fields. However, the body of literature is quite heterogeneous with many small empirical studies consisting of case studies and limited survey research. In recent years there have been increasing attempts to draw conclusions across a larger number of cases, either through meta-analysis techniques applied to several pre-existing case studies or large scale research projects. Furthermore, there is a considerable literature from democracy theory and other normative perspectives on society that consider how the public ought to behave, engage and participate in public decisions. Here too there have been efforts to examine the empirical realities underlying the social goals of normative theories.

Literature Profile

Of the seventy-three (73) references on public perception and participation currently in the literature review, over half are from a resource management context, including 12% from water reuse and biosolids and 10% that are water quality and conservation and watershed management. In fact, many of the water reuse citations are summaries of the research literature from the 1970s and 1980s and thus reflect the vast majority of research on public perception and participation in the water reuse context. Slightly over one-quarter of the literature relate to risk management (e.g., hazardous and nuclear wastes, incinerators, environmental health, etc.). The remaining sources are predominantly land use issues and there are a few general public policy decision-making examples. Nearly 80% are empirical studies (e.g., surveys, case studies, simulations). The remaining are normative theories, emerging from political and social theories of participation in decision-making, communicative action, and justice. The vast majority of examples (94%) are from the U.S., although the literature includes two relevant water studies in Australia, a water reuse survey in England, and two risk management cases from Europe. Overall, the social science disciplines represented in the literature review include:

- 30% Public administration, policy, management, or planning [22]
- 21% Psychology (social and cognitive) [15]
- 16% Political science [12]
- 11% Communication science [8]
- 8% Sociology [6]
- 7% Conflict Resolution [5]
- 6% Marketing and opinion surveying [4]
- 1% Economics [1]

Perception is usually considered an individual-level factor, while participation is a process context, social, and structural-level phenomenon. In other words, perception relates to what is going on in an individual's head -- how they think and feel about things. Participation, on the other hand, refers to all the features of a decision-making process in which individuals are participating. Perception and participation are closely related, as individual factors influence the outcomes, success and characteristics of the process and context for decisions. At the same time, the nature of the process contributes to shaping an individual's perceptions.

The literature contains a wide variety of views on what is called ***success*** in public participation. Some researchers and scholars define success in terms of a broad majority of participants perceiving the process and outcome as fair, based upon sound logic, and legitimate. Others are more concerned that the outcome is implementable or that a particular decision (e.g., water reuse) is acceptable to the public. These two views of success do not necessarily produce the same results in regards to what factors are important in public perception and participation. The perspective of the author is that it is wise to identify both process and outcome dimensions to the success, under the assumption that attending to both simultaneously is the best chance to ensure fair and credible processes and sound, legitimate outcomes. As Caron Chess states in her white paper:

> In general, I think both process and outcome variables need to be considered. I do not feel that good processes necessarily lead to good outcomes. I even think that it is possible to have good outcomes even with what many would consider poor processes

Overarching Themes

In spite of the "messy" relationship among social science factors, a few common themes emerged from literature review. The cross-cutting ideas reflect clusters of factors that influence public participation and perception issues of water reuse. The exact nature of the influence is unclear – it is not possible to say that X causes Y. In fact, there are many chicken-and-egg relationships where it is difficult to decipher whether X leads to Y, Y leads to X, or whether there are several other factors, A, B, and C, that contribute to both X and Y influencing each other.

Nonetheless, general themes are present. When factors within these themes function in certain ways, they tend to have a positive or negative impact on an individual's perception and the nature of public participation. There are at least five themes:

> **Information & Context**: Information, knowledge, local context, education all play an important role in shaping perception and the nature of public participation. The information relates to the science and technologies, local knowledge and site-specific characteristics, values and interests, the local context – e.g., political, social, economic, and environmental landscape, and other information-based factors. Furthermore, the uncertainty or incompleteness of information in any of these information categories influences perception about water reuse. Therefore, information and context factors contain concrete facts, as well as flavor or richness that embellishes the influence of those facts.
>
> **Communication & Dialogue**: Communication is the means of exchanging information, developing mutual understanding and defining relationships. Public dialogue is a broader factor related to the quality of the overall public discussion and cumulative influence of individual communication events. While the effectiveness of the communication events seem to be critical to the success of information exchange and expanding understanding, it is the quality of the overall dialogue that appears to contribute to relationship factors that contribute to public perception and individual's behavior when participating in

public decision-making. For example, people seem to assess communication based upon the:

- sincerity and honesty of the speaker;
- legitimacy and credibility of the speaker;
- comprehensiveness, clarity and coherent (e.g., no jargon or technical language); and
- factual accuracy.[14]

While communication and dialogue are the channels in which information and contextual factors flow, the combined impact of communication and dialogue, and information and context may contribute to shaping the trust and perceived fairness among the individual participants and members of the public.

Trust & Trust-Building: Perception of water reuse and behavior in participation is influenced by trust. Likewise, perceptions impact the level of trust people have in many dimensions of the water reuse issue. It is a cyclical relationship. People can trust or have confidence in many different types of things, including people's trust of technologies, science, the people managing the technologies and systems, and each other. Trust is built through the nature of communication and dialogue and has been shown to increase in cases of high risk perception when:

- dialogue is sustained
- public has independent sources of information, not linked to the sponsoring agency
- public can ask questions
- public is involved early
- information is available to everyone
- behavior is non-coercive – it is considered a reasoned, fair way to make a decision
- everyone's opinion matters and there is a willingness to listen to all views and expand the discussion if necessary, and
- citizens have some level of control in the process (e.g., contributing to the agenda or ground rules).[15]

Fairness: The perception of fairness is important in public participation and perceptions of water reuse. Fairness applies to both the decision-making process and the outcome. For example, was the process of making the decision fair? Was it reasoned, involved everyone, genuinely listened to everyone, addressed all concerns appropriately, etc.?

Fairness also relates to the outcome – that is, are the burdens and/or benefits being shared fairly? A fair outcome may not need to be an equally distributed outcome. From cases of hazardous waste facility siting, it has been shown that people will take on more burdens than others (e.g., one neighborhood will host the landfill), as long as the process of making the decision was perceived as fair *and* there is some level of burden sharing.[16] In other words, while one community takes the landfill, the community sees that a neighboring community is committing to measurable, verifiable waste reduction activities, and a third community is hosting a waste transfer station. Everyone is seen as doing their part.

Motivation & Commitment: Individuals need to be motivated to participate in water reuse decision-making. Rarely is a single motive enough; usually someone needs multiple motives to engage. Commitment, as discussed here, is considered an organizational-level factor in a public dialogue. The organizations involved must show a genuine commitment to public participation that contributes to trust-building and the perceptions of fairness.

These five themes are not independent and distinct from one another. Quite the opposite, they may very likely interact and overlap in multiple ways. In fact, it may be that interaction among the themes that most shapes the public participation experience and the individuals' perceptions of water reuse issues. For example, communication and dialogue seem to be the pipelines through which all forms of information flow. Relationships among people appear closely connected to the information and may be influenced by the effectiveness of the communication channels. The strength of the relationships may contribute to an individual's perception of fairness, the level of trust and degree of trust building (or destruction), and the motivation of individuals and commitment of organizations. It is a complex, circulatory system with factors both influencing and being influenced by other factors. It is difficult to determine where it starts and where it ends, although it is clear when the system fails because conflict escalate and public decisions are derailed (i.e., by the author's definition of success both process and outcome dimensions were ineffective).

IV. What Does All This Tell Us?

Additional lessons can be gleaned from this wide ranging body of literature on public perception and participation. First, a few studies have sought to identify specifically what led to public acceptance of unpopular facilities and less desirable land uses. For example, studies on the public's willingness to accept radioactive waste repositories or hazardous waste facilities have concluded that acceptance may depend upon:

- Reducing uncertainty
- Increasing trust
- Nature and extent of participation opportunities provided
- Knowledge of problem and alternatives
- Perception of credibility and competence of government agency
- Assessment of the technical data
- Burden sharing among communities, including those not hosting the facilities
- Fair decision-making process[17]

A comprehensive study of public participation programs concluded that successful (defined by desirable social goals) public participation processes depended upon:

- Tasks consistent with the capacity and expectations of the public and government
- Open, fair communication, emphasizing deliberation
- High quality, two-way communication among citizens, government and scientists
- Sufficient government resources (financial and staff time) to support the process[18]

Preliminary results from a study of watershed partnerships in California, Oregon and Washington found that there was multiple dimensions of success in partnerships, including: educational and outreach; capacity-building among stakeholders; reaching agreements; project and policy modifications; perceived impacts on the watershed; achievement of goals; and monitoring programs as a proxy for outcomes. Each type of success was associated with a different set of explanatory factors. Nonetheless, trust among stakeholders was the strongest and most consistent factor, followed by age of the partnership, i.e., experience working together.[19]

A few of these studies were some of the first attempts to identify generalizable findings by examining large samples. While they were broad, they could not go into the depth that smaller case studies accomplished and at times they had to consolidate many factors into larger macro-variables to demonstrate statistical significant. Therefore, it is best to interpret these studies in conjunction with the wider literature. The purpose of the literature review is to contribute to our overall understanding of the factors at play in shaping perception and the nature of public participation.

Next Steps

The literature review was the first step in a sequence of events to bring multiple disciplinary and stakeholder perspectives together to clarify the nature, extent, and performance of key factors in water reuse public perception and participation. The next steps include:

- Case summaries from communities that have initiated water reuse projects
- White papers from a panel of experts on public perception and participation
- Symposium in August 2001 to bring together diverse social scientist and water reuse stakeholders for structured, facilitated dialogue
- Framework document that summarizes the entire project, including the literature review, case studies, white papers, and symposium. It will bridge disciplinary and stakeholder perspectives and supplies practical, effective guidance and tools for water resource professionals.

For more information about the *Framework for Public Perception and Participation in Water Reuse Initiatives* project contact the Principal Investigator at RESOLVE, Troy Hartley, Ph.D. (202-944-2300) or the Project Manager at the Water Environment Research Foundation, Bonnie Bailey (703-684-2470).

---

[1] U.S. Environmental Protection Agency, 1992.
[2] Lawrence, 2000.
[3] Jeffery, 2001.
[4] Bruvold, 1988.
[5] U.S. Environmental Protection Agency, 1992; Lohman 1984.
[6] Putnam, Robert D., January 1995, "Bowling Alone: America's Declining Social Capital," *Journal of Democracy*, 6(1):65-78. And, trust in government surveys conducted by The Pew Research Center For The People and The Press (e.g., http://www.people-press.org/questionnaires/trustque.htm).
[7] Lawrence, 2000.
[8] Broad, 1996; Jeffery, 2001.
[9] Lawrence, Dec 2000; Bruvold, 1984; Lohman and Milliken, 1984; Broad, 1996.
[10] Broad, June 1996; Lawrence, 2000; Bruvold, 1984; Lohman and Milliken, 1984.
[11] Lawrence 2000; Broad 1996.
[12] Bruvold, 1981; Bruvold, 1991; Lawrence, 2000.
[13] Lawrence 2000; Broad 1996.
[14] King, Feltey, and Susel, 1998.
[15] Renn, Webler, and Wiedemann, 1995.
[16] Rabe, 1994.
[17] Johnson and Scicchitano, 2000; Kraft and Clary, 1991; Rabe, 1994.
[18] Beierle and Konisky, 1999.
[19] Leach, Pelkey, and Sabatier, 2000; Sabatier, 2001.

ATTACHMENT 1

WHAT IS BEHIND THESE FIVE THEMES? THE DEVIL IS IN THE DETAILS.

The five overarching themes – Information & Context, Communication & Dialogue, Trust & Trust-Building, Fairness, and Motivation & Commitment – reflect clusters of more specific and narrow factors. Each factor seems to contribute to how the theme is expressed in a given case. This section describes the twenty-three factors in greater detail. For our purposes, a factor is any characteristic, phenomenon, or fact that influences in some way the perception held by individuals or the nature of the public participation experience.

For ease of reporting and clarity, the key factors in public perception and participation identified from the literature review are presented below under three general categories: 1) Individual; 2) Decision Context; and 3) Institutional and Societal Context. Individual-level factors relate to the individuals participating directly or indirectly in a public decision making effort (e.g., demographics, knowledge and awareness, motivation). Decision context factors refer to the features of the particular public issue and the mechanisms for public decision-making (e.g., nature of the information available and needed, quality of the dialogue). Finally, institutional and societal factors reflect the broader landscape within which individuals act and decision-making processes are undertaken (e.g., institutions and decision-making authority, organizational culture).

The categories and factors identified here are not completely distinct from one another; rather factors overlap. Factors vary in scale, with a few applying to narrow components of perception and participation (e.g., specific psychological processes) and others to broader elements (e.g., trust). Relative weight and significance for factors are not given, in part because the relationships among factors are not well understood, nor do analytical methods or theoretical frameworks exist to measure factors in relationship to one another. Furthermore, factors likely interact or produce cyclical relationships that make causality difficult to verify. For example, while an individual's attitudes and communicative skills contribute to trust in government officials, technologies and management systems, trust also shapes attitudes and how one chooses to communicate. The reciprocal, feedback impacts of factors on one another produces a chicken and egg type dilemma.

Attachment 1 discusses each factor in more detail; see Table 1 on the following page for a summary of the twenty-three factors. Below is a discussion of the Individual, Decision Context, and Institutional and Societal Level factor categories. Following the description of each factor is a list of literature sources covering the information presented in the description and cross-referenced to the overall reference list in Attachment 2.

TABLE 1: OVERVIEW OF PUBLIC PERCEPTION & PARTICIPATION FACTORS FROM A SOCIAL SCIENCE LITERATURE REVIEW

CATEGORY 1: INDIVIDUAL CHARACTERISTICS	CATEGORY 2: GROUP & DECISION CONTEXT	CATEGORY 3: INSTITUTIONAL & SOCIETAL CONTEXT
• Attitudes • Commitment & Motivation • Communicative Skills • Demographics • Interests, Concerns, & Values • Knowledge & Awareness • Learning Styles • Personal Behavior • Power • Psychological Processes • Risk Perception • Trust	• Complex Information • Dialogue Quality • Empowerment • Fairness • Leadership • Process Features • Representation	• Institutional Authority • Organizational & Professional Cultures • Organizational Commitment • Social Capital

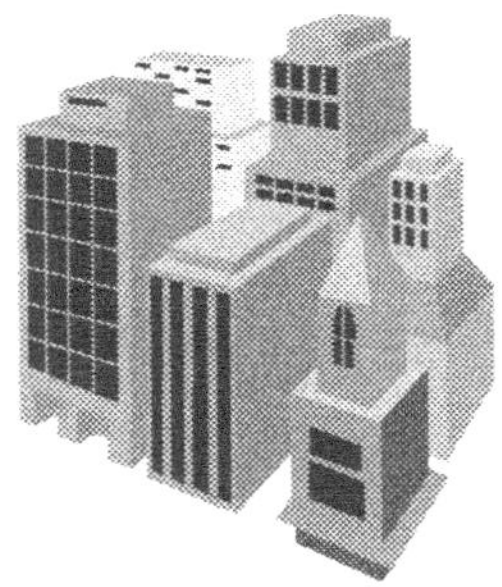

CATEGORY 1: INDIVIDUAL LEVEL FACTORS

Below is a discussion of the individual level factors that the literature identifies as potentially influencing how people think and feel and the nature of public participation.

Attitudes. A set of broad psychological perspectives may contribute to how individuals perceive information, the issues and solutions they believe are relevant, and their roles and responsibilities in decision-making. All individuals, including citizens, elected officials, government officials, interest group representatives, journalists, businesspeople, and others concerned with water reuse are influenced by their attitudes. Attitudes are judgements about or preferences for or against an object, idea, person, institution, or other physical and psychological object.

Attitudes are fairly strongly held, although they can be changed. The strength of an attitude can be measured or indicated by the information the attitude is based upon, the degree of elaboration of the argument in support of the attitude, and its stability and certainty among those holding the attitude. Many factors may contribute to shaping attitudes, including:

- Expertise or level of familiarity with particular issues
- Demographic factors (e.g., race, ethnicity, age, gender, class, ideology, education)
- Social community and natural environment
- Communicative style

Experts seem to view things differently than lay people, regardless of educational level – everyone is an expert in something, having a more detailed knowledge of their area of expertise and experience than others. This is discussed in more detail below in the risk perception section.

Cultural background, for example, may have an impact on how someone frames a problem and whether they view government or experts as the ones responsible for solving the problem. A person's disciplinary training may contribute to their perception of what information and solution options are relevant, and the legitimacy of other's concerns. The social community and natural environment in which someone is raised and/or is living may shape their sense of identity and their perceived role in society. Attitudes may be influenced by communicative style; for example, re-framing is a dispute resolution strategy that seeks to change language and the perception of issues. Changing the frame from "resolving disputes" to "finding constructive improvements in the status quo" may contribute to altering parties' attitudes and perspectives from strongly defending their positions to examining the complexity of an issue and seeking alternatives.

In turn, attitudes may influence the atmosphere of the communication and the quality of the dialogue in the decision-making process. Attitudes may influence how people view decision-makers. Furthermore, these psychological factors may contribute to the most effective learning styles of individuals, (e.g., Native American communities and the role of stories in learning).

Like almost all of the factors identified in the literature, attitudes may exhibit the feedback affects of higher-order factors discussed above. For example, a study concluded that when trust is lacking, if one party frames an issue differently than another, then the second party may attribute the difference to a hidden agenda, narrow self-interests, or moral lapses. Furthermore, if government officials use language and information perceived to be unrelated to frames adopted by the community, the government officials may be judged by the community to be unresponsive and this may undermine trust.

Literature sources: B4; B7; B8; J1; J15; J18; J29; J31; J32; J40; R1; R2; C1; C3.

Commitment & Motivation. Commitment is the willingness and responsiveness to engage in the public decision-making process. Motivation refers to the reasons why people decide to engage and remain engaged in the public issue. The literature identifies a wide range of self-interests and common-interests that motivate participation in public decision-making, including:

- Personal enjoyment, satisfaction, and fun from engagement
- Perception of fairness
- Suspicion of others' interests as opposing one's own interests
- Sense of importance of tasks and impact on community
- Hope for the future and optimism that a good solution can be found
- Framing problems as opportunities

Multiple motives appear to be the most consistent conclusion in the literature. In other words, people may need more than one motive or incentive to act.

Commitment was shown to be particularly important for government officials. Agencies may undermine participation programs if it is not committed to them, engaging in good faith with the intent to follow-through on the outcomes of discussions. Depending upon the nature of the public participation program and the discussions with the public, commitment may require considerable flexibility and willingness to move forward under uncertainty about what comes next in the decision-making process. In highly participatory public discussions, commitment may also require overcoming significant challenges and barriers, e.g., historical relationships and legislative or regulatory requirements.

There is disagreement in the literature over whether common interests emerge from more participatory public deliberations. In other words, some argue that the act of participating in deliberative processes produces more common interest among the participants. Others disagree, believing self-interest remains dominant. Both seem to agree that additional common interests are identified, however they disagree how significant an impact those common interests have on the group.

Literature sources: B1; B4; B6; B7; J4; J7; J10; J13; J20; J21; J22; J23; J25; J26; J29; J30; J32; J34; J42; J44; R1; R7; N1; N2.

Communicative Skills. Communicative skills refer to an individual's abilities to articulate, listen, clarify, reflect upon, and respond to a wide range of types of information (e.g., technical, scientific, values, process, group agreements, legal, etc.). Highly constructive communicators demonstrate many skills, including:

- Listening
- Seeking clarification

- Providing feedback
- Modeling communicative behaviors

Ideas are more effectively communicated by increasing their clarity and making them coherent, increasing its salience -- i.e., relevant to the listener's experiences and knowledge structure. For example, images and simulations, particularly those that include readily identified local landmarks, can be very effective at communicating the outcome of design alternatives. Simple mechanisms of organizing written materials advance understanding by presenting information in way consistent with the configuration of knowledge in the brain, e.g.:

- Overviews for each section
- Headings to break sections of 3-6 paragraphs
- Headings stated as questions and distinguished by a change in print style
- Technical information explained in text, not a glossary
- Bullets to convey technical specifications rather than in the body of text

Communicative skills may interact closely with the type of information being discussed (e.g., the degree of complexity and uncertainty) and the relevance of the information to the problem at hand. In addition, effective communicators model behavior, e.g., an emphasis on deliberation rather than coercion. Furthermore, participants may assess speakers based upon:

- Sincerity and honesty of speaker
- Legitimacy and credibility of speaker
- Thoroughness of information, without use of jargon and technical language
- Factual accuracy of information.

The importance of communicative systems and the act of communicating are themes that cross the categories of factors. Effective communication may play a crucial role in public perception and participation at all levels.

Literature sources: B6; J7; J22; R5; R7

Demographics. A wide range of demographic factors may influence certain perceptions and behaviors, in part because they reflect deeper psychological and social belief systems. Relevant demographic factors from the literature include:

- Age
- Income
- **Culture or ethnicity**
- Education
- Gender
- Ideology (i.e., political party, interest group affiliation)
- Length of residency
- **Race**

Culture, ethnicity, and race seem to have a more powerful influence than other demographic factors. Communication is filtered through belief and value systems, and cultural, ethnic, and racial experiences and norms may contribute to the structure and content of the belief and value systems. In certain cases (e.g., low-income, African American and Native American) the belief and values systems may lead to a higher degree of mistrust and suspicion in government. Length of residency, as a person interacts daily with the natural and social features of their community, has been linked to a person's beliefs and values system through establishing their sense of place and identity. Thus, a perceived risk may be viewed as more threatening to a long-time resident in the water reuse context, although other studies in different context have suggested the opposite.

In general, demographic factors may influence:

- Perception toward water reuse, hazards, and other scientific and technical issues
- Commitment and willingness to participate (including likelihood of contacting political representatives)
- Interests and values
- Trust
- Process and outcome expectations
- Behavioral norms

However, there appears to be important modifying effects as demographic factors interact with contextual factors. For example, while surveys in California and Colorado have shown older women to be less supportive of potable water reuse and hazardous waste siting in part because of perceived public health concerns and the general interpretation that women are more risk averse than men, a study in risk perception raises the possibility of an important exception. In a New Jersey study where communities were already stressed with multiple hazards (e.g., landfills, hazardous waste sites, chemical and industrial facilities, airports, blighted buildings, crime, etc.) both men and women respond similarly to perceived risks.[20] Furthermore, another study from the risk perception literature suggested that *together* age, education, gender, societal risk aversion, personal risk aversion, perceived appropriateness of actions, perceived detail of action, and technical details of action accounted for a relatively small percentage of variation in perceived risk and a smaller percentage of variance in behavioral intentions.[21] In other words, while these demographic factors may allude to whether or not someone would perceive risks from a hazardous facilities and was willing to do something about it, many other factors not studied and not well understood may contribute significantly to those perceptions and behaviors. While research on demographics produce a cloudy picture, one things seems clear -- relying on demographics alone will not provide an adequate picture of public perception and participation.

Literature sources:B1; J6; J16; J35; J41; J42; R2; R8; C1

Interests, Concerns, & Values. Individual's interests, concerns and the expression of values relate to a specific context. Interests and concerns are the individuals' articulated reasons for believing or acting. They may include a set of desires, needs, and expectations that exist in a particular decision context. Values are deeper, core principles, desires, and expectations that are often reflected in the interests and concerns expressed by people. Consequently, differences over values are much more difficult to overcome than differences in interests or concerns. In fact, the negotiation research has shown benefits from re-framing disputes and conflicts from differences in concrete positions and values to a discussion over each parties' interests and concerns. While the underlying value differences may not be resolved, both parties' interests and concerns sometimes can be adequately addressed in "win-win" solutions that permit constructive improvements in a status quo.

Research within a water reuse context has identified several specific public interests, concerns and values. The following were discussed in the introductory section of this literature summary; however, they included:

- Public health and safety concerns
- Cost, capability and reliability of treatment and distribution systems concerns
- Water conservation and environmental protection values
- Water supply independence interests

In general, interests, concerns, and values have been shown to be shaped by many factors, including demographic characteristics, and knowledge and awareness levels. Knowledge and awareness includes recognition of one's interests, including economic and other value interests, awareness of risks, and other context-specific impacts on a person's values and interests. Interest, concerns, and values can vary by scale. For example, if an individual represents a local interest group, their interests, concerns, and values may differ from the same interest group's regional, state, or national representative.

While interests, concerns, and the expression of values are issue-specific, there are a few broader values that have been observed across cases, e.g.:

- Democratic values
- Justice

Some scholars argue that the public has certain democratic values. Many also believe that access to clean water is a fundamental right. Further, they believe the public holds an expectation that they should serve a role in decision-making and be kept informed of issues. When that role is not provided, and particularly if the opportunity is taken away without their knowledge and consent – and they find out – then significant suspicion and mistrust may be generated. Likewise, there is an expectation for justice. These democracy scholars believe the public's interest is in a fair decision-making process, equity in the distribution of the outcome (both burdens and benefits), and a non-coercive approach by the more powerful participants.

Fairness and procedural justice has received attention in the literature, and in particular, research has considered what participants' perceive as fair in a public decision-making process on controversial environmental and natural resource issues. A discussion of fairness exists below in the Decision Context Level Factors, since it is a value and interest particular to the decision-making process.

Literature sources: B3; J3; J21; J23; J24; J31; J38; J41; J42; R1; R2; R4; R7; R8; C3; M2.

Knowledge & Awareness. Knowledge and awareness relate to the understanding and application of information relevant to the particular decision at hand. Generally, the public is not scientifically trained and may have less of an understanding of specific scientific and technical issues under consideration in many complex public decisions. While education level attainment is one measure of knowledge and awareness, there are many types of relevant information, including technical and scientific information, decision-making procedures, community interests, values and dynamics, and the nature of the problem. Distinctions are made in the literature between generalized expert knowledge and site-specific local knowledge. Both can have relevance and importance to decision-making; a challenge for decision-making is to effective information sharing and communication between those that posses expert knowledge and those versed in local knowledge.

In one study, a distinction was made between knowledge (i.e., knowing something) and creativity (i.e. using what one knows). There may be an application dimension to knowledge and awareness that is important in public decision-making.

The research on diffusion of suggests that there will be less public support for an innovation when there is uncertainty in the scientific community and competing scientific views expressed in the media and in public dialogue.

The psychology literature has identified important distinctions between knowledge and behavior. In fact, there is little support for the notion that educating the public, increasing their awareness of relevant issues, and adjusting their attitude or values toward the issue will result in the intended behavior. Rather, individuals need knowledge *and* intent before they will act. Intent is an attitude and motivation related to a specific behavior. In the suggested causal link between knowledge and behavior, intent is an intermediary factor and is controlled in part by a range of factors:

- Motivation
- Interests and concerns
- Convenience of the desired behavior
- Perception

Awareness of water reuse in areas of the country where reuse initiatives exist is fairly low, often well less than half of a surveyed population sample. In one survey, short explanations of the water reuse process increased support for indirect potable reuse, although it also intensified the opinions of those opposing it. Public opinion surveys over the past three decades have found the respondents with higher levels of educational attainment more supportive of the indirect potable reuse idea than others. However, when examining opinions surrounding specific, real-life cases of reuse in a respondent's community rather than the general concepts, support wanes among the highly educated. This is consistent with the literature on the NIMBY (Not In My Back Yard) phenomenon in siting waste facilities and other undesirable land uses in a community. In NIMBY cases, opponents have been found to be highly educated on the specifics of the technologies and the science involved in proposed uses.

Literature sources: B1; B2; B6; J4; J5; J18; J23; J26; J29; J43; J44; R1; R2; R5; R7; R8; R8; C3; M2

Learning Styles. Learning is closely related to knowledge and awareness, since it is the process of converting information to knowledge and awareness. While learning styles may be a basic, first-order factor, there are more impacts of learning styles than simply absorbing information. On the basic level, every individual has his or her own optimal learning styles. Different cultural backgrounds may predict individual learning styles (e.g., Native American communities often employ stories as learning techniques). Consequently, processes that employ multiple techniques of communicating and presenting ideas can be more effective at producing a knowledgeable public and participants.

Furthermore, people are most effective learners when information is explained in a holistic and systematic manner, which is consistent with intuitive perspectives. Systems thinking considers the interrelatedness of forces and elements, viewing them as part of a common process. People are more effective learners if they are permitted to guide and influence the types of information to be learned. For example, people may be more motivated learners when they have a role in identifying the information and knowledge needed to make a decision. This takes time and requires patience.

In addition, learning may have broader impacts on issues of public perception and participation, including links to:

- Motivation
- Mutual understanding, respect and social capital
- Coping

Learning is linked to motivation, since knowledge and awareness can increase public interest and involvement. In one study, sharing information was found to be motivational. In addition, based upon survey results in California, many people appear to want to learn and have expressed an interest in learning about water reuse and participating in the decisions about water reuse. Many dispute resolution and sociology scholars believe a group of diverse stakeholders learning together may build mutual understanding and social capital (i.e., the societal norms, networks, and social trust that permit coordinated action). While group learning efforts have built the norms, networks and trust of social capital, there is disagreement whether it fosters a group identity with common interests. Finally, confidence in one's ability to learn may be important when tackling unfamiliar, complex information. In other words, promoting learning and an individual's experience with learning complex information may help individuals cope.

Literature sources: B2; B6; B7; J7; J8; J10; J23; J29; J31; J35; J39; J44; R1; R5; R7

Personal Behavior. An individual's behavioral patterns in public interactions either promote or interfere with several civic values identified by normative democracy theories. In particular, behavioral factors that shape public perception and participation include:

- Mutual respect, including respecting others individually and their constraints
- Honesty
- Constructive interaction and personal behavior
- Fear of trying or engaging in public dialogue
- Ability to balance multiple roles in public dialogue

One study found a government official, who was perceived by the public as more personable than his predecessor, far better able to serve multiple roles as a regulator, educator, and innovator than his predecessor. Other research found citizens wanting constructive personal interactions in public participation processes. In the end, behavior does matter.

Literature sources: B6; J29; J34; J38; R5; C6

Power. Power is a broad factor that includes many elements that supply an individual with control and influence. It can be derived from many sources, including:

- Financial resources
- Human resources
- Decision-making authority
- Coalition building and maintenance
- Technical and scientific expertise and/or access to technical and scientific assistance

Research has shown that financial resources and access to technical assistance alone do not make citizen participation effective. In addition, the capacity to build coalitions that can support positions is needed. In forms of participation that involve negotiation among stakeholders, there is a need for on-going communication with constituencies of representatives. This ensures that the constituency groups support the agreements negotiated by their representatives.

Research on technical assistance for community groups has only begun to illuminated what technical assistance can and cannot accomplish. For example, technical assistance to organized citizen advisory groups may:

- Promote monitoring, environmental reviews and analyses
- Contribute to environmental improvements
- Improve the quality of decisions

Technical assistance may not:

- Promote citizens' acceptance of a technical decision or siting location
- Promote political consensus
- Bolster citizens clout in decision-making

Power is not merely an inherent social characteristic, it is also defined by its use. In other words, how someone applies the power that they have may make a difference in public perception and participation. This is important in part because power inequities remain and can emerge in any public participation process. In many participatory processes, administrators have been found to be most successful when they serve multiple roles, balancing their authority and power as stakeholders, information sources, and interpreters of legal requirements.

Authority-sharing may empower participants. However, recent research on public participation programs suggest that flexibility and genuine responsiveness of decision-makers to the concerns and issues raised by others may be more important than or as effective as formal power-sharing arrangements. In fact, the act of participation may generate a sense that government is more responsive and listens. In other words, in some cases of successful public participation, government may maintain authority while sharing power.

Literature sources: B1; B6; J5; J6; J8; J11; J21; J22; J26; J41; J44; R1; R7; C6; P1.

Psychological Processes. Psychological processes are patterns in human's cognitive processing. Specific processes that impact public decision-making situations include:

- Jumping to conclusions
- Bias toward the familiar
- Limited attention capacity
- First impressions establish the frame of mind
- Coping

People are quick to jump to conclusions and apply stereotypes. Jumping to conclusions is predicting outcomes based upon prior knowledge and experience. It provides greater predictability and makes it easier to function, particularly under conditions of uncertainty and/or complexity. For example, negative relationships between parties are quite durable and familiar, and parties will have selective perceptions of each others' actions which reinforce negative impressions of one another.

Similarly, people have a strong bias toward the familiar and will distort information that runs counter to their sense of the familiar. People will look at a new situation or problem, quickly label it as similar or the same as a previously solved problem, and then apply the solution to the earlier problem in the new context. The bias is so strong that if information in the new situation contradicts someone's perception of the context as familiar, people will exclude or ignore the contradictory information and not perceive the competing facts. Strongly held, familiar ways of doing things or views of the world are hard to break.

People's attention capacity is more limited than the informational resources available for decision-making. People will cognitively fatigue if managing too much information, uncertainty, or unfamiliar information. When fatigued the biases are stronger and the individual's learning and cognitive effectiveness is inhibited. Cognitive restoration will occur with time, a break, and other restorative activities.

The initial representation of a problem, or framing of the issue, is critical to how the psychological biases will function. First impressions do matter, in part because the initial representation triggers whether and how the psychological processes function.

A narrow psychological factor, coping, was identified as potentially important in the public policy and administration fields for its impact on successful public deliberation among multiple participants with diverse backgrounds and interests. Coping is a stable pattern of cognition and behavior that helps people function in the face of uncertainty, confusion, discomfort, and other sources of stress. While psychologists have studied coping for years, they have not often examined it in an environmental or public decision-making context. A study discovered that government officials' patience can be strained by citizens and political activists. Likewise, citizens and community leaders can also exhibit patience and have their patience strained during their interactions within government decision-making processes. The need to cope may arise from many characteristics of public participation programs and interactions between the general public and highly technical and scientific experts. In particular frustration may arise from the:

- Lengthy, exhausting participation processes
- Behavioral challenges
- Hard ball negotiating tactics and strategic behavior
- Conflicts
- Misunderstandings and miscommunication

Coping lessens the consequences of these frustrations. Coping strategies that have been seen in some cases and may be helpful include:

- Versatility
- Patience
- Non-Defensiveness (including a sense of humor)

Other types of factors, such as motivation and learning, have been identified as possibly serving coping functions as well.

Literature sources: B6; B7; J1; J10; J28; J31; J34; J5; R5

Risk Perception Risk perception has been studied predominantly within the context of hazards (e.g., hazardous waste management, nuclear waste management, occupational and environmental health). There is fairly strong support for the notion that laypersons and experts do not share the same perspective on the sources or severity of risks. The general public may be more concerned with the impacts and severity of an accident than the likelihood of an accident, whether the risk is voluntarily or involuntarily imposed upon them, and the difference of opinions that exists within the technical and scientific communities. The expert, on the other hand, appear to be less concerned with the severity of risk and the difference of opinions among scientists, attending more to the risk probabilities and relative risks.

Existing knowledge structures and experiences may be important in risk perception. People do not interpret risk information in isolation, but rather it seems to be filtered and interpreted in the context of their existing knowledge structure. So, ignoring what people already know and believe in risk communication may confuse people and fail to inform them.

While some demographic characteristics may correlate to certain risk perceptions (e.g., income levels are linked to willingness to accept greater occupational risks, females are generally more risk averse than males), recent studies have cast doubt on the power of these trends. In one study, multiple hazards in a neighborhood (e.g., blighted communities with several less desirable land uses, high crime, drugs, etc.) were found to eliminate some gender differences and was hypothesized to alter variation in risk perception by age, education level, economic class, risk aversion, and other demographic factors.

There have been studies examining the role of the media in shaping risk perception. For example, a simulation study found that if the media frames issues with distrust and controversy, it seems to increase the perception of risk, even if the factual risk information is held constant.[22]

Literature sources: J12; J16; J17; J20; J23; J36; J38.

Trust. Trust emerges consistently in the literature as a key factor at play in perception and public participation. Trust is a large factor with many components that relates to the belief and confidence in the reliability, credibility, and fairness of people, organizations, decision-making processes, management systems, technologies, or other dimensions of decision-making and implementation. Trust affects individuals differently. In general, trust in government is in decline.

The public seems to trust and believe in science and technology for the most part, and they believe that adequate analysis is needed prior to making a decision. However, they may not always trust the sources of information, including government, industry, and the media. The public seems to trust university scientists and the medical community as more credible than other sources of information. Nonetheless, adherence to expert knowledge at the expense and disregard for local knowledge and concern seems to lead to mistrust and suspicion among local communities, particularly when the expert knowledge is inconsistent with local knowledge or perspectives. In such cases of inconsistency between expert and local knowledge, the motives of the expert or scientist may be called into question and their information may be discarded.

Public trust in government may be promoted when:

- Sustained dialogue exists
- Participants are able to secure independent expert advice
- Participants are free to question others
- Early involvement in a public decision
- All information readily available to all involved
- Logical and transparent process of selecting options
- Decision-makers take seriously or endorse the outcome of a public participation process
- Citizens have some level of control of the format of decisions (e.g., agenda, rules)

There is overlap in what builds and promotes trust and what people perceive as a fair process. Consequently, perception of fairness may relate closely to whether the participants and processes are viewed as trustworthy – acting fairly may contribute to making someone trustworthy.

Literature sources: B1; B4; B7; J20; J22; J23; J25; J34; J38; J44; R1; R8; M2.

CATEGORY 2: DECISION-MAKING & SITUATIONAL CONTEXT LEVEL FACTORS

Below is a discussion of the decision-making and situational context factors that the literature identifies as potentially influencing how people think and feel and the nature of public participation.

Complex Information. Information includes the diverse types of data, evidence, values and other information that relates to the individuals involved, specific problem and decision-making process at hand, and the broader institutions in which decisions are made and implemented. The information itself is communicated, learned, and generally managed in a manner shaped by the many factors discussed in this literature review.

Information has a complexity dimension that is comprised of many dimensions, including:

- Degree of scientific uncertainty
- Degree of scientific disagreement
- Unique perspectives on particular pieces of information held by specific groups (e.g., culture, race, ethnicity, expert, layperson)
- Characteristics that make information less universally understood and harder to communicate

The variety of information types calls for a variety of communicative, learning, and management strategies.

Recently concerns have been raised by members of the scientific community (e.g., U.S. EPA's Science Advisory Board) about whether participatory, stakeholder involvement mechanisms promote the level of dialogue necessary to ensure a high-quality, science-based decision. In other words, some are concerned that quality science is getting short-changed in favor of public values and uninformed decisions.

The U.S. EPA Science Advisory Board undertook an investigation of the issue and while the report is currently being finalized they likely will conclude that participatory processes that involve consultation and negotiation among involved stakeholders can result in high-quality, science-based decisions (see draft final report on Science Advisory Board's web site, www.epa.gov/sab). This may be particularly true when decisions include:

- Careful and complete review and analysis of available scientific evidence
- State-of-the-art methods of assessing the available evidence
- Integration of scientific understanding, communities' interests and values, and government agency's interest in the broader public interest
- Stakeholders identify and agree on whom should be responsible for obtaining needed factual information

Literature sources: J4; J6; J7; J18; J19; J20; J26; J31; J32; J40; R1; R4; R7; R8: R9; C3; M1.

Dialogue Quality. The quality of a public dialogue is a set of characteristics that contribute to how constructive and productive discussions are among participants within the broader context of political society. It is closely related to process, information, and communication factors discussed elsewhere, although it may also integrates these ideas and shapes the relationships among participants. There is a wide range of issues and types of information that must be effectively communicated, including:

- Technical and scientific information
- Legal and decision making procedures
- Interests and values
- Emerging agreements

According to normative communicative action theories and a few empirical studies, a communicative system that produces a high quality dialogue and promotes mutual understanding, trust, and social capital may include, at least:

- Face-to-face discussion
- Frequent interactions
- Simple, accurate and comprehensive explanations

The simple, accurate and comprehensive explanations appears to create rich imagery, i.e., a holistic cognitive image of a decision or outcome that has relevance to past experiences and existing knowledge.

Different kinds of information may often need different presentation and communication strategies because of the nature of the information and people's diverse learning styles. Communication systems that promote dialogue through multiple venues may be most effective at dealing with these differences. There are many approaches to managing and sharing information, including technical assistance, computer mapping and models, media, etc.

Different groups or types of participants (e.g., citizens, elected officials, and government managers and experts) sometimes communicate as if they speak different languages. In other words, each group may articulate their interests, issues, and problem solving perspectives in unique ways that may not be readily understood by members of the other participants groups. A study of hazardous waste regulation in several states that examined these communicative perspectives concluded that there are very few opportunities in society to promote dialogue across these groups speaking different "languages."

In public participation processes that depend upon representatives from interest groups, constituencies may fail to achieve the same level of mutual understanding, trust, and social capital that developed among the representatives. If this is true, it would place significant demands on the representative to consistently inform their own constituency and maintain their role as a legitimate voice for their constituency. The constituency may need to be sufficiently well informed and supportive of their representative to endorse the decision. (See the discussion on representation under Process Features above)

Finally, the public outreach, information and education approaches that utilize the media as a predominant communication channel may not always be conducive to promoting a high quality dialogue, since little discussion and multi-way communication can take place. There are some examples of media-sponsored series of public forums that promote interactive, sustained public dialogue. Furthermore, while studies have found that the public does not generally trust the media as a source of accurate information, the media is the predominant information source for most of the public.

Literature sources: B2; B3; B7; J10; J17; J18; J19; J22; J27; J32; J36; J39; J43; J44; R1; R8; R9; R8; P1; M1; N3.

Empowerment. Empowerment is a higher order factor that is complex and has many components. It involves increasing a person's capacity to define, analyze, and act upon his or her own problems. It includes individuals who possess personal capacity to influence decisions, and decision-making process and organizations that build and promote personal capacity. It has emerged in social science theory over the past couple of decades to understand complex individual, community, and societal interactions among:

- Self-esteem
- Self-efficiency
- Knowledge and skills
- Political awareness
- Social participation
- Political participation
- Rights and responsibilities
- Access to psychological, social and material resources

Empowerment at the individual level has three dimensions: 1) intrapersonal – perceived personal capacity to influence social and political systems; 2) interactional – knowledge and skills to master social and political systems; and 3) behavioral – actions that influence social and political systems. There are four forms of empowerment, all of which are necessary for community empowerment, although different forms may be achieved without achieving others:

- Formal
- Intrapersonal
- Instrumental
- Substantive

Formal empowerment is created when institutions (governments, businesses) provide mechanisms for the public to influence decisions. Intrapersonal empowerment is a feeling of personal competence in a given situation. Instrumental empowerment refers to the individuals' actual capacity for participating in and influencing a decision-making process. Substantive empowerment refers to the ability to reach decisions that solve problems or produce desired outcomes and it requires that citizens and formal institutions work together to reach decisions.

In the environmental decision-making context, processes that are not empowering may present perceived threats to public health, disrupt existing social networks, and provide a loss of personal control. Dis-empowerment can be extremely frustrating and may lead to a "reactive" form of empowerment where groups mobilize their resources and skills to respond to the perceived threat. In fact, the mobilization and organization of a response can be empowering for the participants, supplying procedural, psychological, and substantive satisfaction. An increase in empowerment without adjusting the administrative system (e.g., genuine commitment, power-sharing, real opportunity to shape outcome, new roles that change experts to partners) may also lead to frustration.

Literature sources:B1; B5; J5; J8; J9; J10; J22; J27; J33; R1

Fairness. While the concept of fairness has permeated throughout other discussions of factors, it is sufficiently important to warrant its own discussion here. Perception of fairness relates to both outcomes (i.e., fairly distributed burdens and benefits) and the process of making the decisions. Being treated fairly communicates respect, that one's views are important, and that the person is a valued member of the community. The public's perception of fairness often depends upon:

- Government's receptiveness to citizen's input
- Influence citizens have over decision-making process and outcome (procedural justice)
- Quality of government's knowledge and reasoning
- Burden sharing of outcome (distributive equity)
- Degree to which relationships improved during the process

Government's receptiveness and citizens' influence over decisions relate, at least in part, to the degree to which the public had control over the process and decision, including a role in problem identification and finding a solution. Control includes being given opportunities to present evidence and voice their opinions. In other words, citizens want to be heard in a meaningful way, know that their opinion matters, and be treated with respect. Furthermore, government responsiveness and citizen control relate, in part, to the consistency with which procedures are applied over time and across individuals, and the neutrality and trustworthiness of decision-makers. The quality of the knowledge and reasoning concerns the use of accurate information, the existence of opportunities to correct or modify decisions, and the representation of important perspectives in the process. In addition, the quality of the reasoning can include the maintenance of ethical standards throughout the decision-making process.

While the perception of procedural fairness may result in satisfaction with the decision process, fairness also relates to the decision outcome. In some cases, there seems to be an expectation that decisions will reflect a sharing across society of the negative, undesirable results. In a U.S.-Canadian comparative case study on hazardous waste siting, the researcher found that the burdens do not necessarily need to be equal; however, everyone seems to have to demonstrate they are doing their part and contributing to addressing the problem and issue. In this case, a community was willing to accept a hazardous waste landfill, in part because they saw that others outside of their community who benefit from the landfill were taking on some responsibility for overall hazardous waste management, e.g., hosting a transfer station, committing to concrete waste reduction measures.

However, a "frustration effect" has been observed in situations where participants perceived procedural fairness, but still were unsatisfied in the decision process and/or outcome. While the frustration effect is not well understood or studied, some scholars suggest it occurs when communities feel dis-empowered, particularly when two conditions may exist:

- negatively perceived outcome that is exacerbated by repeated disappointment with the decision-makers or broader social support for the perception that the outcome is unfair; *AND*
- decision-maker perceived as having a personal stake in the outcome, rather than being viewed as an impartial judge.

Frustration effects have not been seen in studies of legal settings, with perceived neutral judges making decisions.

Other research has concluded that risk management agencies often encounter this frustration effect because they frame the public issues (e.g., hazardous waste facility decisions) as factual arguments about risk probabilities and the extent of potential harm. Meanwhile, citizens and communities are framing the issues as more intense concerns about the institutional competencies to deal with risks, and/or different social or cultural values. Nonetheless, communities are forced to make their arguments about their value concerns on factual grounds of risk probabilities and human health risks, resulting in the perception that their views are not heard, not respected, and do not matter. For the government official, these arguments are perceived as irrational.

The perception of procedural fairness differs from citizen to citizen, although the relative importance of procedural fairness may relate to:

- Gender (males less concerned with fairness than females)
- Favorability of decision to one's interest
- Quality of outcome (e.g., science-based, perceived as well-reasoned)
- Level of conflict

Literature sources: J24; J25; J31; C6.

Leadership. A fairly specific, narrow set of factors relate to the role of a leader and/or policy professional in shaping the decision-making context, promoting an empowering atmosphere by setting the tone and expectations of the decision-making process, advancing innovation, and reaching consensus among diverse interest groups. Leaders may function as collaborative leaders, leveraging the skills of individual members of a group, facilitating group processes, and articulating agreements. They often serve roles that are viewed as legitimate by a large number of interest groups, and are not necessarily the participating individual with the most authority or power.

Policy professionals are staff and managers in agencies and implementing organizations across local, regional, state, and national levels who share a common interest in a particular policy approach or outcome. They form networks of like-minded officials and build the networks' capacity to promote a particular policy innovation, such as water reuse. Policy professionals can serve champion roles for an innovation.

In successful participatory processes, government agencies have demonstrated leadership in the following areas:

- Ensuring the quality of the communication
- Clarifying the scope of tasks
- Providing resources
- Ensuring the quality of the process
- Demonstrating a commitment to legitimate public involvement.

Literature sources:B3; J2; J5; J21; J22; R1.

Process Features. Process features are the specific design characteristics of the public participation program and activities. There are many options for decision-makers, each with strengths and weaknesses, and each producing different kinds of outcomes. What is considered a success in a public participation process can be defined in many ways. Some objectives relate to pragmatic, constructive progress on the decision at hand, while others involve broader societal goals related to democracy and individual rights. For example, the literature includes the following success measures:

Outcome Success

- Acceptable decision
- Consensus, reduce conflict
- Educated public
- Improved quality of decision
- Incorporate public values
- Builds trust
- Build capacity of participants and the group to work together in the future

Process Success

- Fairness
- Information exchange
- Access to information
- Procedures adhered to by group
- Group processes
- Representation

However, many factors beyond the structure of the participation process account for the variation in public participation success seen in the literature. For example, studies on citizens' perceptions of a good, successful participation process have identified several elements:

- Access to process, fairness
- Power to influence the process and outcome, balanced use of power
- Access to necessary information
- Structural characteristics promote constructive interactions
- Facilitation of constructive personal behaviors
- Adequate analysis, information and reasoning; not political or power-driven reasoning
- Enabling of future processes through capacity-building, social capital

Many of these factors relate to the individual, contextual and institutional/societal factors discussed elsewhere in this literature review. So, the manner in which a process is designed can influence the formation of many of the characteristics and factors perceived as important by citizens, but process design alone may not necessarily ensure a successful public participation program.

Under certain circumstances that are not well understood, participation may:

- Defuse hostility
- Increase tolerance of differences and patience in participation
- Build trust and sense of responsiveness of government
- Increase mutual understanding
- Increase appreciation of the limits of government
- Increase confidence of citizens to participate
- Increase knowledge of political systems and case-specific information
- Increase motivation to participate

At the same time, it appears that participation may or may not:

- Overcome deep-rooted social and community problems (e.g., socioeconomic and racial biases)
- Result in representative participation – often the processes under-represent disadvantaged groups and favor advantaged groups within communities
- Smooth the path of implementation of decision
- Increase the number of participants (particularly in the case of more open, deliberative processes)

There is disagreement over whether participation builds a sense of community. Some argue that it does, while others have found little empirical support for it. There may exist agreement that a greater sense of community and common identity may emerge from public participation programs. However, the question of significance of that sense of community remains unresolved.

Public participation programs in the literature faced several distinct challenges, including but not limited to:

- Integrating science and technology adequately can be difficult
- Processes can be difficult to manage
- Stakeholders not well defined
- Processes can be intimidating to citizens
- Organizations may not be committed to it (e.g., inadequate funding and staffing)
- Social and political history may have hardened resistance

Which form of public participation selected (e.g., public meetings, workshops, advisory groups, citizen panels, deliberative polls, town hall meetings, charettes, etc.) may depend upon many issues, including the level and nature of the conflict or disagreements over the decision at hand. For example, some scholars suggest that if the disagreement is over facts, then education may be the primary objective and process design feature. If there are distinctly different social values lifestyles at play, then extensive information sharing and co-learning may prove helpful. In today's complex public decision-making context, often education, trust building, and value differences are only three of many issues that must be addressed in a public participation program. Table 2 on the next page summarizes many of the characteristics of different forms of public participation, including factors that enhance or limit their effectiveness.

Overall, successful public participation programs are often expensive, slow and challenging. Several characteristics may enhance their likelihood of success, including:

- Staff and financial commitment
- Liaison to promote participation
- Neutral, competent facilitator
- Government commitment to follow recommendations
- Clarify public participation goals, since they are likely to vary
- Clarify roles of participants
- Collect feedback on public participation effort
- Design a process that can accommodate conflict and resolve disputes
- Avoid condescension toward any participant, no matter what their limitation

Recent studies have shown that a wide range of processes can be successful at achieving several public participation goals and objectives. So, it may not be the form of the process that is critical, but rather the dialogue itself. Processes must promote dialogue. Nevertheless, opportunities for face-to-face dialogue alone may not be sufficient; there may also be a need for government endorsement and commitment. The communication must be open, fair, and 2-way among the participants.

Literature sources: B1; B3; B6; J2; J4; J7; J8; J9; J11; J13; J14; J26; J28; J38; J40; R1; R3; R4; R8; B1; M1.

[1] Greenberg, 1995.
[1] Sandman, et al., 1993.
[1] Sandman, et al., 1993.

TABLE 2: PUBLIC PARTICIPATION DESIGN OPTIONS*

Types	Some of the Possible Process & Outcome Features	Characteristics that *MAY* Limit Effectiveness	Characteristics that *MAY* Enhance Effectiveness
Public Meetings	• Focal point of opposition • Less likely to yield consensus • Can lead to subsequent institutional changes, organizational & social learning	• Inadequate Agency outreach • Limited provisions of technical information • Procedures that disempower citizens • Unwillingness to accommodate discussion of social issues • Timing of hearings: held late in decision-making process	• Holding meetings in combination with other forums • Provide significant technical assistance • Conduct vigorous outreach • Encourage participation of native peoples (and other under-represented groups) • Discuss social issues • Field Questions adequately.
Workshops	• Process perceived as fair • Improve understanding of issues	• Inadequate outreach • Logistics: held at inconvenient time or location	• Site-visits, experiential activities • Joint fact-finding among participants • Holistic, systems thinking approach to information sharing • Connections to existing knowledge structure and experiences • Multiple presentation techniques
Citizen Advisory Groups	• The weaker the ties to the sponsoring Agency, the less controlling the Agency might be perceived, but the less likely they might be to deal with significant issues. • Scope may include influence on the development of risk assessments and constructive changes to facility siting plans. • May permit citizens monitor environmental reviews and analyses, and secure project improvements • May improve quality of decision • May not promote citizen acceptance or promote political consensus	• Agency overly controlling • Value judgements • Institutional or structural separation from planning process	• Outreach to variety of channels for support of recommendations • Professional members (possibly). • Composition of committee (representative) and credibility of members; mechanisms of accountability to broader community • Utilize pre-existing citizen networks and infrastructure • Committed members • Input sought early in decision-making process • Agency has genuine interest in the input; independence of advisory group from sponsor • Political commitment • Clear tasks, agendas, expectations and roles • Facilitators • Use of information sources beyond sponsor • Focus on issues of significance to participants

* Every public participation process is unique and not all types of designs will have all of the features identified in Table 2. Furthermore, there is insufficient empirical work to be certain about the factors that limit or enhance effectiveness. Therefore it is important to view the public participation design features in Table 2 not as definitive and independent processes and factors, but rather as general rules of thumb to guide practitioners – if particular processes or efforts to enhance their effectiveness do not work, be willing and ready to try something else.

There are many variations on public participation mechanisms and new mechanisms that have not been well studied. For example,

- Citizen Panels (Juries) – assign randomly selected citizens to hear testimony from experts and make a decision.
- Deliberative Polls – randomly select a large sample, supply them with intensive background information and briefings over 2-3 day period, and then conduct a poll to obtain an "informed citizen's" view.
- Town Hall Meetings – moderated large group meeting with citizens asking representatives or officials questions.
- Charrettes – use of rough models of a design or plan that small groups of citizens manipulate and discuss, arriving at an agreed upon design.

(see, Renn, Webler, and Wiedemann, 1995; Fishkin, 1995).

Representation. While representation can be considered one of the process features, the nature of representation has drawn considerable attention in the literature. It is controversial and reflects a debated area in the literature; therefore, it is discussed here as its own factor.

Some suggest that what constitutes a stakeholder is unclear and thus makes selection of representatives for participatory processes difficult. For example, since the concerns of the same interest group can vary along a local, regional, state and national scale, identifying the most appropriate representative can be difficult.

Others add that many of the highly participatory processes (e.g., stakeholder negotiations) do not permit large numbers of the community to participate; they must remain relatively small to effectively negotiate and reach agreements. This may violate democracy values and principles that seek to prevent control by the elite and promote active citizenry.

Still others argue that different parts of the decision-making process (e.g., problem definition, option generation, analysis, implementation) require different stakeholders and representatives.

Furthermore, representatives may need adequate skills to function and represent their constituency's interests effectively and to communicate and maintain accountability with the broader public. In some cases, research has found that citizens may lack skills in coalition building, constituency maintenance, negotiation, communication, collaborative multi-party problem-solving, and group processes. In one study, surveyed representatives of interest groups were mixed in their beliefs of whether citizens could be trusted to be valuable participants in participatory processes.

Socially and economically disenfranchised groups have historically been under-represented in many participatory processes. In part because of these implementation difficulties and the normative, democratic and justice concerns, some argue that interest group representative processes should be used sparingly in favor of randomly selected citizen panels. Citizens act as juries, hearing evidence from the interested parties, and making decisions. Some proponents of citizen panels argue that the random selection process to identify representatives for a citizen panel is preferable to interest group representatives because randomly selected citizens are uncommitted to a position, and can take advantage of new information and adjust their preferences accordingly. In addition, randomly selected citizens may be less concerned about the social status and power of each individual member outside of the panel, since they may be less likely to interact with the stakeholder representatives in the future.

Literature sources: B1; J8; J26; J27; J31; J32; J37; J43; J44; R1; R8; P1; M1; C4

CATEGORY 3: INSTITUTIONAL & SOCIETAL CONTEXT FACTORS

Below is a discussion of the institutional and societal context factors that the literature identifies as potentially influencing how people think and feel and the nature of public participation.

Institutional Authority. Overall, institutional and societal factors relate to the cultural and behavioral norms that shape the societal context in which individuals and decision-making processes take place. Decision-making authority is a basic component and it is established through law and regulation. Specifically, who is the ultimate decision-maker and what are the legal administrative procedural requirements for decision-making?

In fact, one of the challenges for decision-makers in contentious public policy decisions may be how to manage the authority in such a way that both retains the authority and shares the power, since promoting the trust-building, empowerment, and commitment discussed above depends in part on power-sharing through public involvement. Administrative procedural requirements may constrain or may provide opportunities for the deliberation and effective management of authority and power-sharing. There appears to be a communicative dimension to managing authority to retain it and share power at the same time. For example, a study found that clearly articulating 1) a genuine interest in implementing the outcome of a participatory process to the fullest extent possible, 2) a commitment to negotiate in good faith, *and* 3) that the government agency is the ultimate decision-maker seemed effective for government officials in a participatory public decision-making process. It appeared to retain and ratify the agency's authority, and it seemed to contribute to building trust, empowerment, and commitment needed to achieve an acceptable, implementable decision.

Literature sources:J9; J12; P1.

Organizational & Professional Cultures. Organizations and disciplinary professions have their own culture (and sometimes more than one) -- i.e., a persistent, patterned way of thinking about tasks and relationships to other people within and outside of their organization and/or profession. Cultures are strongly held and resistant to change. Tasks and relationships outside the norms of an organizational or professional culture are often difficult to accomplish effectively or efficiently.

The norms and expectations determined by organizational and professional cultures contribute to many of the factors discussed above. For example, they may contribute to the specific assumptions and conclusions that experts and laypersons reach quickly as a result of the psychological processes, such as jumping to conclusions. Cultures may shape the nature and expectations of communication styles, and the distinct "languages" spoken by community members, elected officials, government bureaucrats, attorneys, scientific and technical experts, and other distinct organizational or professional groups.

Literature sources:J9; J12; P1.

Organizational Commitment. While it has been mentioned within the context of other factors, an organizational commitment to public involvement seems to be important. In other words, genuine commitment that communicates good faith may be different than the perception of public involvement as "window dressing," as a citizen called it in one study. Organizational commitment, as distinct from an individual's commitment as a representative of an organization, suggests that there are broader components to demonstrating commitment, e.g., resources, staffing, consistent messages, and others.

Literature sources: R5

Social Capital. Social capital is the societal norms, networks, and social trust that permit coordinated action. Social capital is the atmosphere and overall social environment in which decision-making takes place. Social capital has emerged as a social science concept in the theory-building around political systems and democracy, in particular.

Scholars suggest that social capital can support cooperation and civic engagement for mutual beneficial gains, or if built around different societal objectives (e.g., totalitarianism), it may support rigid, power-driven, coercive decision-making. Social capital is simultaneously a product of the individual interactions and decision-making processes that define the behavioral norms, social networks, levels of trust *and* a strong influencing factor on what behaviors, interactions, and relationships are possible in a society.

Social capital can reside in neighborhood strength, a sense of community and belonging, and the mutual respect people show toward one another. It appears that through deliberation and learning together social capital may be built to produce an atmosphere conducive to agreements. However, as has been suggested above, social capital may also exist that inhibits high quality public dialogues. This is particularly true in cases where the underlying norms, networks and social trust define an atmosphere of mistrust, distinct and opposing interests, and limited deliberation among citizens, interest group leaders, government and elected officials, businesspeople, and other participants. Since social capital is a relatively new theoretical concept, it is not clear how to modify it. It would appear to be resistant to change in a similar manner that cultures are resistant to change.

Literature sources:B1; B2; J7; J18; J40; R1; R7.

CONCLUSION

The twenty-three factors above reflect a summary of those variables discussed in a broad array of social science literature. The specific literature sources referenced at the end of each discussion of a factor are listed in Attachment 2.

ATTACHMENT 2

REFERENCE LIST

BOOKS:

B.1. Berry, Jeffrey M., Kent E. Portney, and Ken Thomson, 1993, *The Rebirth of Urban Democracy*, Washington DC: The Brookings Institution

B.2. Fishkin, James S., 1995, *The Voice of the People: Public Opinion & Democracy*, New Haven CT: Yale University Press

B.3. Rabe, Barry G., 1994, *Beyond NIMBY: Hazardous Waste Siting in Canada and the United States*, Washington DC: The Brookings Institution

B.4. Renn, Ortwin, Thomas Webler, Peter Wiedemann, 1995, *Fairness and Competence in Citizen Participation: Evaluating Models for Environmental Discourse*, Boston MA: Kluwer Academic Publishers

B.5. Schwerin, Edward W., 1995, *Mediation, Citizen Empowerment, and Transformational Politics*, Westport CT: Praeger.

B.6. Thomas, John Clayton, 1995, Public Participation in Public Decisions: New skills and strategies for public managers, San Francisco CA: Jossey-Bass

B.7. Williams, Bruce A., and Albert R. Matheny, 1995, *Democracy, Dialogue, and Environmental Disputes: The Contested Languages of Social Regulations*, New Haven CT: Yale University Press

B.8. Wondolleck, Julia M. and Steven L. Yaffee, 2000, *Making Collaboration Work: Lessons from Innovation in Natural Resource Management*, Washington DC: Island Press

JOURNAL ARTICLES:

J.1. Bardwell, Lisa V., 1991, "Problem-Framing: A Perspective on Environmental Problem-Solving," *Environmental Management*, 15(5):603-612.

J.2. Bens, Charles, January 1994, "Effective Citizen Involvement: How to Make it Happen," *National Civic Review*, pg.32-39.

J.3. Bruvold, William H., January 1988, "Public opinion on water reuse options," *Journal of Water Pollution Control Federation*, 60(1):45-50.

J.4. Burroughs, Richard, December 1999, "When Stakeholders Choose: Process, Knowledge, and Motivation in Water Quality Decisions," *Society & Natural Resources*, 12(8):797-809

J.5. Busenberg, George J., September 2000, "Resources, Political Support, and Citizen Participation in Environmental Policy," A Reexamination of Conventional Wisdom," *Society & Natural Resources*, 13(6):579-587.

J.6. Cantrill, James G., 1998, "The Environmental Self and a Sense of Place: Communication Foundations for Regional Ecosystem Management," *Journal of Applied Communication Research*, 26(1998):301-318.

J.7. Chess, Caron and Kristen Purcell, August 1999, "Public Participation and the Environment: Do We Know What Works?" *Environmental Science & Technology*, 33(16):2685-2692.

J.8. Cohen, Nevin, June 1995, "Technical Assistance for Citizen Participation: A Case Study of New York City's Environmental Planning Process," *American Review of Public Administration*, 25(2):119-135.

J.9. Curtis, Allan and Michael Lockwood, January-February 2001, "Landcare and Catchment Management in Australia: Lessons for State-Sponsored Community Participation," *Society & Natural Resources*, 13(1):61-73.

J.10. Daniels, Steven E., and Gregg B. Walker, 1996, "Collaborative Learning: Improving Public Deliberation in Ecosystem-Based Management," *Environmental Impact Assessment Review*, 16(1996):71-102.

J.11. DeHaven-Smith, Lance, and John R. Wodraska, July/August 1996, "Consensus-Building for Integrated Resource Planning," *Public Administration Review*, 56(4):367-371.

J.12. Flynn, James, Paul Slovic, and C.K. Mertz, 1993, "Decidedly Different: Expert and Public Views of Risks from a Radioactive Waste Repository," *Risk Analysis*, 13(6):643-647.

J.13. Frey, Bruno S., and Felix Oberholzer-Gee, 1996, "Fair Siting Procedures: An Empirical Analysis of Their Importance and Characteristics," *Journal of Policy Analysis and Management*, 15(3):353-376.

J.14. Gericke, Kevin L., and Jay Sullivan, 1994, "Public Participation and Appeals of Forest Service Plans – An Empirical Examination," *Society and Natural Resources*, 7():125-135

J.15. Gregory, Robin, June 2000, "Using Stakeholder Values to Make Smarter Environmental Decisions," *Environment*, 24(5):34-44.

J.16. Greenberg, Michael R., and Dona F. Schneider, August 1995, "Gender Differences in Risk Perception: Effects Differ in Stressed vs. Non-Stressed Environments," *Risk Analysis*, 15(4):503-511.

J.17. Halfacre, Angela C., Albert R. Matheny, and Walter A. Rosenbaum, 2000, "Regulating Contested Local Hazards: Is Constructive Dialogue Possible Among Participants in Community Risk Management," *Policy Studies Journal*, 28(3):648-667.

J.18. Hisschemoller, Matthijs, and Cees J. H. Midden, January 1999, "Improving the usability of research on the public perception of science and technology for policy-making," *Public Understanding of Science*, 8(1):17-33

J.19. Innes, Judith E., Winter 1998, "Information in Communicative Planning," *Journal of the American Planning Association*, 64(1):52-63.

J.20. Johnson, Renee J., and Michael J. Scicchitano, 2000, "Uncertainty, Risk, Trust, and Information: Public Perceptions of Environmental Issues and Willingness to Take Action," *Policy Studies Journal*, 28(3):633-647.

J.21. Kathlene, Lyn and John A. Martin, 1991, "Enhancing Citizen Participation: Panel Designs, Perspectives, and Policy Formation," *Journal of Policy Analysis and Management*, 10(1):46-63.

J.22. King, Cheryl Simrell, Kathryn M. Feltey, and Bridget O'Neill Susel, July/August 1998, "The Question of Participation: Toward Authentic Public Participation in Public Administration," *Public Administration Review*, 58(4):317-326.

J.23. Kraft, Michael E., and Bruce B. Clary, June 1991, "Citizen Participation and the NIMBY Syndrome: Public Response to Radioactive Waste Disposal," *The Western Political Quarterly*, 44(June 1991):299-328.

J.24. Lauber, T. Bruce and Barbara A. Knuth, January/February 1999, "Measuring Fairness in Citizen Participation: A Case Study of Moose Management,: *Society & Natural Resources*, 11(1):19-37

J.25. Lawrence, Rick L., Steven E. Daniels, and George H. Stankey, November-December 1997, "Procedural Justice and Public Involvment in Natural Resource Decision Making," *Society & Natural Resources*, 10(6):577-589.

J.26. Lynn, Frances M., and George J. Busenberg, April 1995, "Citizen Advisory Committees and Environmental Policy: What We Know, What's Left to Discover," *Risk Analysis*, 15(2):147-162.

J.27. McCay, Bonnie J. and Svein Jentoft, 1996, "From the Bottom Up: Participatory Issues in Fisheries Management," *Society & Natural Resources*, 9(1996):237-250.

J.28. McKenzie-Mohr, Doug, May 2000, "Fostering Sustainable Behavior Through Community-Based Social Marketing," *American Psychologist*, 55(5):531-537.

J.29. Michaels, Sarah, Robert J. Mason, and William D. Solecki, January 1999, "Motivations in Ecostewardship Partnerships: Examples from the Adirondack Park," *Land Use Policy*, 16(1):1-9.

J.30. Plumblee, John P., Jay D. Starling, and Kenneth W. Kramer, February 1985, "Citizen Participation in Water Quality Planning: A Case Study of Perceived Failure," *Administration & Society*, 16(4):455-473.

J.31. Renn, Ortwin, Thomas Webler, and Branden B. Johnson, Summer 1991, "Public Participation in Hazard Management: The Use of Citizens Panels in the U.S.," *Risk – Issues in Health & Safety*, 197(Summer 1991):197-226.

J.32. Rich, Richard C., Michael Edelstein, William K. Hallman, and Abraham H. Wandersman, 1995, "Citizen Participation and Empowerment: The Case of Local Environmental Hazards," *American Journal of Community Psychology*, 23(5):657-676.

J.33. Roberts, Nancy, March/April 1997, "Public Deliberation: An Alternative Approach to Crafting Policy and Setting Direction," *Public Administration Review*, 57(2):124-132.

J.34. Rosell, Ellen, 1996, "The Sunland Park/Camino Real Partnerships: Landfill Politics in a Border Community," *Policy Studies Journal*, 24(1):111-422.

J.35. Sandman, Peter M., Paul M. Miller, Branden B. Johnson, and Neil D. Weinstein, 1993, "Agency Communication, Community Outrage, and Perception of Risk: Three Simulation Experiments," *Risk Analysis*, 13(6):585-598.

J.36. Sellers, M.P., 1993, "NIMBY: A case study in conflict politics," *Public Administration Quarterly*, 16(1993):460-477.

J.37. Slovic, Peter, M.Layman, N.Kraus, James Flynn, J. Chalmers, and G. Gessell, 1991, "Perceived Risk, Stigma, and Potential Economic Impacts of a High-Level Nuclear Waste Repository in Nevada," *Risk Analysis*, 11(4):638-696.

J.38. Sullivan, William C., Frances E. Kuo, and Mona Prabhu, 1997, "Communicating with Citizens: The Power of Photosimulations and Simple Editing," *Environmental Impact Assessment Review*, 17(1997):295-310.

J.39. Tuler, Seth, and Thomas Webler, July/August 1999, "Voices from the Forest: What Participants Expect of a Public Participation Process," *Society & Natural Resrouces*, 12(5):437-453.

J.40. Vaughan, Elaine, 1995, "The Significance of Socioeconomic and Ethnic Diversity for the Risk Communication Process," *Risk Analysis*, 15(2):169-180.

J.41. Verba, Sidney, Kay Lehman Schlozman, and Henry E. Brady, May/June 1997, "The Big Tilt: Participatory Inequality in America," *The American Prospect*, 32(May/June 1997):74-80

J.42. Weeks, Edward C., July/August 2000, "The Practice of Deliberative Democracy: Results from Four Large-Scale Trials," *Public Administration Review*, 60(4):360-372.

J.43. Wondolleck, Julia M., Nancy J. Manring, and James E. Crowfoot, 1996, "Teetering at the Top of the Ladder: The Experience of Citizen Group Participants in Alternative Dispute Resolution Processes," *Sociological Perspectives*, 39(2):249-262.

J.44. Yearly, Steven, April 2000, "Making systematic sense of public discontents with expert knowledge: two analytical approaches and a case study," *Public Understanding of Science*, 9(2):105-122

REPORTS:

R.1. Beierle, Thomas C., and David M. Konisky, September 1999, *Public Participation in Environmental Planning in the Great Lakes Region*, Washington DC: Resources for the Future.

R.2. Broad, Paul A., June 1996, *Community Views on Water Reuse*, Sydney, Australia: Sydney Water Corp, Ltd.

R.3. Bruvold, William H., August 1991, Public Evaluation of Municipal Water Reuse Options: Technical Completion Report, Berkeley CA: University of California.

R.4. Charnley, Gail, July 2000, Democratic Science: Enhancing the Role of Science in Stakeholder-Based Risk Management Decision-Making, Washington DC: HealthRisk Strategies.

R.5. Hartley, Troy, 1998, *Participatory Competencies in Deliberative Discourse: Cases of Long-term Collaborative Environmental Decision-Making*, unpublished dissertation, Ann Arbor, MI: University of Michigan.

R.6. Jeffery, Paul, 2001, *Understanding public receptivity issues regarding 'in-house' water recycling: results from a UK survey*, unpublished manuscript. Cranfield University, UK.

R.7. Lawrence, Gary, December 2000, *Survey of Orange County Voters Within The Orange County Water District*, Santa Ana CA: Lawrence Research – Public Affairs & Marketing.

R.8. Daisey, Joan, and Granger Morgan, October 7, 1999, Science Advisory Board Commentary on the Role of Science in "New Approaches" to Environmental Decisionmaking that Focus on Stakeholder Involvement, EPA-SAB-EC-COM-00-002, Washington DC: U.S. EPA Science Advisory Board.

R.9. Yosie, Terry F. and Timothy D. Herbst, September 1998, Using Stakeholder Processes in Environmental Decisionmaking: An Evaluation of Lessons Learned, Key Issues, and Future Challenges, Washington DC: Ruder Finn.

CONFERENCE PAPERS:

C.1. Bruvold, William H., 1984, "Obtaining Public Support for Innovative Reuse Projects," Proceedings of the Water Reuse Symposium III: Future of Water Reuse, Volume 1, San Diego, CA, August 26-31, 1984, pg. 122-131.

C.2. Leach, William D., Neil W. Pelkey, and Paul A. Sabatier, November 2000, "Keys to Success in Watershed Management Partnerships: Approach and Initial Results," Paper deliverd to the 8th Biennial Conference of the Watershed Management Council, November 27-30, 2000, Pacific Grove, CA.

C.3. Lohman, Loretta C., and J. Gordon Milliken, August 1984, "Public Attitudes Toward Potable Wastewater Reuse: A Longitudinal Case Study," Proceedings of the Water Reuse Symposium III: Future of Water Reuse, Volume 1, San Diego, CA, August 26-31, 1984, pg. 109-121.

C.4. McCloskey, Michael, May 18, 2000, "Guidance for Environmental Advocates Considering Use of Collaborative Processes," Comments at the Conference on Alternative Dispute Resolution and Natural Resources: Building Consensus and Resolving Conflicts in the Twenty-First Century, U.S. Institute for Environmental Conflict Resolution: Tucson, AZ.

C.5. Sabatier, Paul, March 7, 2001, "Seven Measures of Success for Watershed Partnerships and an Analysis of the Factors Affecting Success," Partners' Seminar Series, RESOLVE, Inc., Washington, DC.

C.6. Timney, Mary M., 1996, "Overcoming NIMBY: Using Citizen Participation Effectively," Paper presented at the 57th National Conference of the American Society for Public Administration, Atlanta, GA, June 29 – July 3, 1996.

RESEARCH PROPOSALS:

P.1. The National Research Council, Commission on Behavioral and Social Sciences and Education, Division on Social and Economic Studies, August 2000, "Public Participation in Environmental Assessment and Decision-Making," Washington DC: The National Research Council. NAS Proposal No. 00-CBSSE-140-01.

MANUALS:

M.1. Bryan, Todd A., January 1997, Effective Stakeholder Involvement Strategies for Environmental Technology Projects: Principles and Practical Applications, Denver CO: Colorado Center for Environmental Management.

M.2. U.S. Environmental Protection Agency, September 1992, *Manual: Guidelines for Water Reuse*, Office of Water, and Office of Research and Development, EPA/625/R-92/004.

NEWSLETTERS & NEWSPAPER ARTICLES:

N.1. Babcock, Thomas M. and Jane H. Ploeser, August 1998, "From Expert Model to Citizenship Model: Phoenix Revises Its Approach to Water Conservation Planning," *PA Times*, 21(8):1, 3.

N.2. Bruvold, William H., February 1981, "Public Participation in Environmental Decisions: Water Reuse," *Public Affairs Report: Bulletin of the Institute of Governmental Studies*, University of California – Berkeley, CA, 22(1):1-6.

N.3. Williams, Michael Paul, October 25, 2000, "Sustained dialogue called key to accord," *Richmond Times-Dispatch*, Richmond, VA, pg.B5.

APPENDIX B

PUBLIC PERCEPTION & PARTICIPATION IN WATER REUSE:

CASE SUMMARIES

JULY 2001

Troy W. Hartley, Ph.D.

TABLE OF CONTENTS

Table of Contents i

I. Case Overview B-1
- Introduction
- Methods
- Case Features

II. Gwinnett County, Georgia B-4

III. City of San Antonio, Texas B-15

IV. City of San Diego, California B-22

I. CASE OVERVIEW

INTRODUCTION

Water reuse, reclaimed water, recycled water, repurified water are all terms for the beneficial use of water after it has been treated in a state-of-the-art, multi-stage wastewater treatment process. Worldwide, water is reused for industrial purposes, landscaping, agricultural irrigation, and indirect and direct potable (i.e., drinking) uses. In the U.S., reclaimed water has been used predominantly for non-potable (i.e., non-drinking) purposes which is widely accepted and utilized, while there are only a few indirect potable examples. As early as the 1930s, reclaimed water was discharged on to the ground and allowed to recharge the groundwater aquifer that serves as a drinking water supply in California. In the late 1950s and early 1960s, additional indirect potable reuse projects in California and Nevada were initiated, as was a temporary, drought-driven direct potable reuse in Kansas – i.e., reclaimed water being directly routed to a drinking water treatment plant. By the mid-1970s, Virginia had an indirect potable reuse project. The 1980s and 1990s were a period of rapid expansion of water reuse, as communities in Colorado, California, Texas, Florida, and Arizona considered indirect potable uses. The 2000s promise additional non-potable and indirect potable initiatives, particularly as population and economic growth pressures further tighten water supplies. However, at the same time public has become increasingly uncomfortable and less accepting of indirect potable projects.[1]

Three case examples are attached on the water reuse experiences in Gwinnett County, Georgia, San Antonio, Texas, and San Diego, California. They are illustrations of the types of issues that water reuse initiatives face, and are intended to give symposium participants a story in which they can relate their own experiences and expertise.

METHODS

Cases were selected based upon the following criteria:

- U.S. potable and non-potable reuse examples – both perceived successes and failures;
- Cases with public involvement and community relations program;
- Multiple stakeholders involved in decision;
- Geographic representation;
- Contemporary cases (last 10 years); and
- Multiple scales (large, medium, and small scale cases).

For each case, documentation was collected using web-based resources, including case documentation on web sites of local, state, and federal governments, local and state environmental organizations, local citizen, business and community groups and associations, and local newspaper archives. Interview questions were designed based upon the outcome of the documentation review and guided by the factors identified in the literature review. Additional documentation was provided by some interviewees, e.g., master plan documentation, public involvement plans, newspaper articles, etc. Draft case summaries were circulated to an Advisory Group and the Project Subcommittee for review and comments. Based upon the comments and

[11] U.S. Environmental Protection Agency, September 1992, *Manual: Guidelines for Water Reuse*, Office of Water and Office of Research and Development, EPA/625/R-92/004.

gaps identified in our review of the case write-ups, additional questions were developed and additional interviews were conducted and analyzed. Additional documentation was gathered from web sources and from those interviewed. The following participant types were interviewed in the cases:

Gwinnett County	San Antonio	San Diego
• County government official • County elected official • Local journalist • Citizen advisory board member • Community group representative (opponent) • Environmental group representative	• City government official #1 • City government official #2 • City contractor	• City government official • State elected official • City council official • Citizen group representative • Environmental group representative • Expert advisory board member

Interviews ranged from a single 30-60 minute phone call to a series of phone discussions over several days.

CASE FEATURES

Several of the case features are contained in the chart on the following page, highlighting differences and similarities across cases. After the chart, the following three cases are summarized:

- Gwinnett County, GA
- City of San Antonio, TX
- City of San Diego, CA

Case Comparison Chart

CASE	Content	PROPOSED SOLUTION	CITIZEN RESPONSE	ADVISORY COMMITTEE	OTHER PUBLIC PARTICIPATION	MEDIA RELATIONS	CURRENT STATUS
San Antonio	• Water shortage and rationing • Lawsuit over environmental impact of water use • Legislation mandating planning • Reservoir voted down • Recycling used in area for years	• Variety of measures and criteria for selection • Recycling to commercial users	• Research showed recycling concerns re cost and creditability • Positive news coverage • Environmental justice concerns over possible residential reuse	• Considering water policy, broad and diverse • Proposed measures and criteria	• Outreach	• Little interest	• Construction well underway, • general support
Georgia	• Population growth • Lawsuits over allocation by Fla. and Ala. • History of sewage spills	• Waste water master plan • Advanced tech plant • Nonpotable use • Use of Lanier for discharge	• Support for initial plant • Opposition of expansion from business/resident association on lake and environmental groups	• Facilitated by GA county Dept. of Public Utilities • Reps from 7 stakeholder groups • Consensus on supply, but dissention over waste water • Long term membership	• Input into water plan • Facilitation by county (with some problems) • Honesty in admitting mistakes	• Characterized as positive by government officials	• Expansion underway • Appeal of state permit by environmental groups & lake association • Oversight by citizen advisory board • Spills and fire
San Diego	• Use of imported water • Pilot project for reliability and safety • Settlement of lawsuit resulted in potable water reuse system	• Use extra reclaimed water in water supply (Repurified Water Project)	• Editorial opposition (toilet to tap) • Environmental justice concerns over use of reservoir for reclaimed water • Cost concerns	• Expert panel • Broad stakeholder review committee • NRC review • Blue ribbon panel	• Updating of city council (different perspectives between government and city council) • Public hearing • Presentations	• Characterized as antagonistic by government officials.	• Despite support of blue ribbon panel, council committee voted to end project • Goal of 25% reuse Currently 3% reuse

II. GWINNETT COUNTY, GEORGIA: F. WAYNE HILL WATER RESOURCE CENTER

Abstract:

Rapid population and economic growth, an extended drought, and water allocation negotiations with neighboring Florida and Alabama have all contributed to Gwinnett County, Georgia's need to reassess its water strategy. While building an 11-step treatment facility called the North Advanced Water Reclamation Facility (NAWRF), the County undertook a multi-stakeholder planning process to update their *Water and Wastewater Master Plan*. The NAWRF was a $245M, 20 million gallons/day (mgd) plant that produced effluent ten times cleaner than traditional wastewater treatment plants, although opponents believe it cannot maintain this level of effectiveness over time. It is located in the north portion of the county on over 700 wooded acres between two highway interstates. A 20-mile pipeline was built to discharge into the Chattahoohee River. NAWRF began operation in late 2000 and was renamed the F. Wayne Hill Water Resource Center for the County Commissioner, who championed the project. A $330M expansion of the plant was approved to increase its capacity to 60 mgd. A second pipeline of 10-miles will be constructed to discharge the effluent in Lake Lanier, the drinking water supply for Gwinnett County and others in the Atlanta metropolitan area. While the state has permitted the expansion and Lake Lanier discharge, the permit is under appeal by state and local environmental groups and an association of businesses, homeowners, and recreational users on Lake Lanier. The appellants believe the plant's discharge will harm the lake and are not convinced that maintenance will be a priority over the long-term. As an alternative to discharge into the lake, some opponents are recommending blending the effluent with water in a reservoir and transferring it to the drinking water plant for a more direct for of indirect potable reuse. However, direct potable reuse is prohibited under state law.

INTRODUCTION: WHY WATER RECLAMATION IS AN ISSUE IN GWINNETT COUNTY

Gwinnett County is a fast-growing county northeast of Atlanta, Georgia. Growing by over 25,000 people/year, its population is estimated to double by 2050 to over 1 million residents. Gwinnett County contains the Eastern Subcontinental Divide, which shunts precipitation on the north side to the Chattahoochee River and ultimately to the Gulf of Mexico, and precipitation on the south side flows into tributaries of Oconee and Ocmulgee Rivers and then to the Atlantic Ocean. The county receives 50 inches of precipitation annually and 36 inches per year during a 100-year drought stage. The U.S. Army Corps of Engineers built a dam on the Chattahoochee and created the 38,000-acre Lake Lanier in 1957. Lake Lanier and the Chattahoochee River and its tributaries supply drinking water for a ten-county region in north Georgia, including the City of Atlanta. While drawing water from the Chattahoochee River basin, including Lake Lanier, counties were discharging treated wastewater on both sides of the Eastern Subcontinental Divide; transfers from one river basin to another is being discouraged by the State Environmental Protection Division (EPD), as is further discharge into the Chattahoochee. (see map on next page)

Wastewater treatment plants in the area, and particularly in the City of Atlanta, were notorious for sub-standard treatment and accidental releases – "sewage spills and overflows, fish kills and a foul-smelling Chattahoochee River have become part of an alarming norm."[2] Furthermore, Gwinnett County had failed to locate a new wastewater treatment plant on the south end of the

[2] Anderson, Will, September 30, 1997, "Site in preparation for county's high-tech treatment facility." *The Atlanta Journal and Constitution*, pg. 01J.

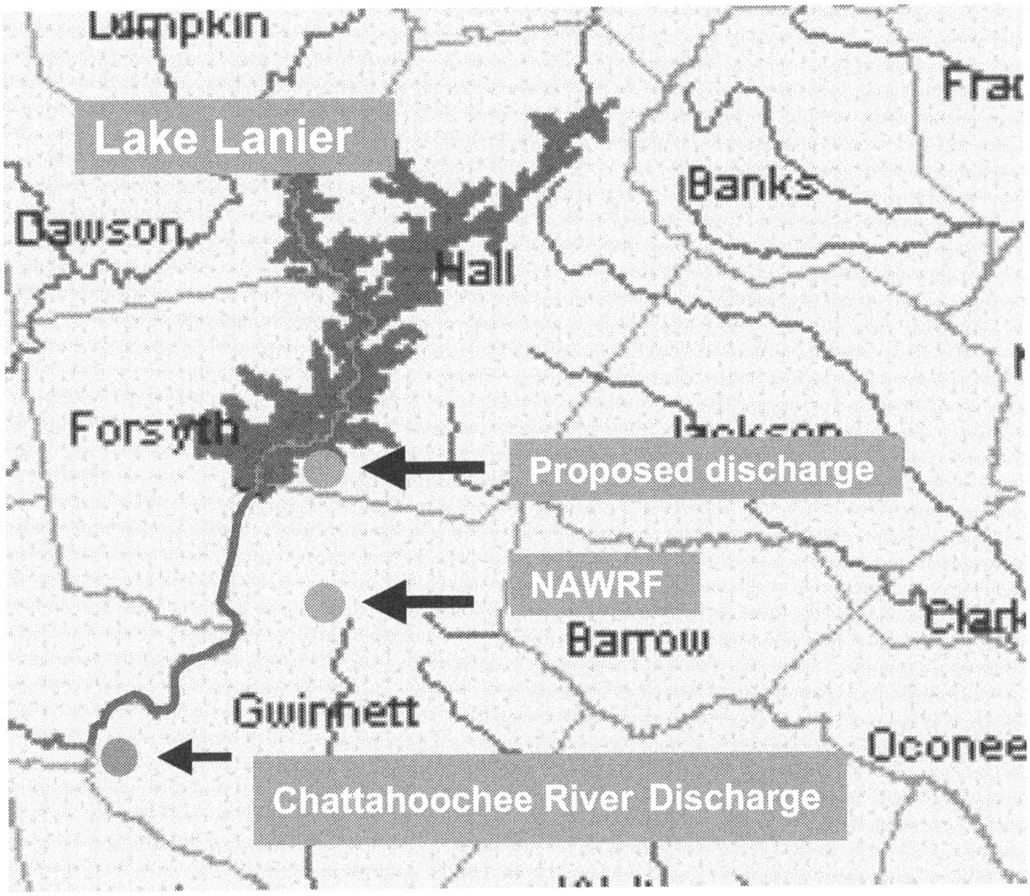

county in 1995, after encountering intense public opposition.[3] Residents had sued the County over its treatment plant on Yellow River/Sweetwater Creek, resulting in a 1990 settlement in which the county bought homes. In 1995, Gwinnett also experienced a lightening-induced sewage spill of more than 500,000 gallons into Norris Lake.[4] (see timeline on page B-14)

How Gwinnett County Responded to Its Water Constraints

In the midst of an extended drought and with state-level negotiations underway on a lawsuit filed by neighboring Alabama and Florida over water allocation from the shared river basins, Gwinnett County updated its water and wastewater strategy in the 1990s.

[3] Matlock, Glenn, staff writer, March 14, 1996, *Gwinnett Daily Post*, untitled emailed article.
[4] Anderson, Will, September 30, 1997, "Site in preparation for county's high-tech treatment facility." *The Atlanta Journal and Constitution*, pg. 01J.

North Advanced Water Reclamation Facility (NAWRF)

One of Gwinnett County's responses to their wastewater treatment needs was building the North Advanced Water Reclamation Facility (NAWRF), approved by the county in 1995. The County Commissioner, F. Wayne Hill, identified a 700-acre site between two Interstate highways as a potential site. The nearest residents were more than a mile away. The NAWRF covers 120-acres, leaving the remaining woods and wetlands for a planned nature center.

However, in part due to the reluctance of the State EPD to permit more discharge into the Chattahoochee River, Gwinnett proposed an advanced 11-step reclamation facility that included primary, secondary, and advanced treatment, and a 20-mile pipeline to discharge into the Chattahoochee. The state issued a draft permit in 1996. Construction began in 1998, and the facility was dedicated in December 2000. As one county official characterized it, NAWRF is "a treatment plant with two drinking water plants attached to it" including:[5]

- Clarifying tanks
- Biological treatment
- Membrane filters
- Sand and activated carbon filters
- Ozone gas disinfection

Officials claimed they wanted to "set a new standard in the Southeast" and return water to the Chattahoochee that is cleaner than what currently exists there.[6] The plant was modeled after the most advanced water treatment plants in the country and "will remove 10 times more pollutants from waste water than a typical" wastewater treatment plant.[7] However, a group opposed to the discharge plans at the plant believes, "the technology is no different...it's like a typical sewage plant. State of the art, they say, ...but it's really no different...just small refinements."[8]

According to one county official, an average 20 million gallons/day plant would cost approximately $60M. The NAWRF cost $275M, $30M of which was for the 20-mile long, oversized discharge pipeline. The pipeline would permit non-potable reuse customers to tap the line for reclaimed water on its way to the Chattahoochee River. The costs included $60-80M spent on back-up systems for the pipes, tanks, electrical systems, and fiber optics in the computer operations center -- "reliability through redundancy," said one county official.[9] The project is being financed through bond funds and public utility resources, including a water rate increase in 2001. While the average rates for Atlanta metro counties is $48-50/month for 7000gallons, Gwinnett county's rates are $51/month.[10]

---

[5] Interview with county official, May 2001.

[6] Brack, Elliott, March 28, 1997, "Hill a fitting site for high-concept treatment plant," *The Atlanta Journal and Constitution*, pg. 02J. And, Matlock, Glenn, staff writer, August 28, 1996, *Gwinnett Daily Post*, untitled emailed article.

[7] Shelton, Stacy, April 29, 1998, "Borrowing big for new water treatment plant," *The Atlanta Journal and Constitution*, pg. 01JJ.

[8] Interview with Lake Lanier Association official, July 2001.

[9] Matlock, Glenn, staff writer, December 3, 2000, *Gwinnett Daily Post*, untitled emailed article. Meyers, Susan Laccetti, November 1, 2000, "Waste Water: Gwinnett will keep on flushing," *The Atlanta Journal and Constitution*, pg. 18A. Interview with a county official, May 2001.

[10] Interview with county official, May 2001.

NAWRF was supported by many in the region. Local environmentalists were cautiously optimistic; "I would like to see every wastewater treatment plant as capable as the North plant. But then anything manmade can have problems. Gwinnett County is going to have to prove through plant maintenance and adherence to the state permit levels that they can set a higher standard."[11] As early as 1997, environmentalists were saying "We'd like this plant to set a precedent for the technology that all sewage treatment plants should obtain."[12]

Opponents expressed concern that the plant would promote further growth, and was being supported by the development community. As one state lobbyist said, "water is becoming more important than roads."[13] A state official in considering a proposed Gwinnett County discharge permit noted that the county could have to slow growth, "don't let development outstrip your infrastructure."[14] Others opponents have expressed concern over the effectiveness of long-term maintenance, fearing that upkeep with get a low priority, saying you "cannot trade it in after three years like a car." They cite the three spills and one fire at the facility as an example, "there already are spills, so how good is it?"[15]

WATER & WASTEWATER MASTER PLAN

In late 1996, Gwinnett County began to update their existing Water and Wastewater Master Plan. It was completed in the summer of 1998, after "extensive citizen input to develop a plan that would be flexible and into which resident's viewpoints could be incorporated."[16] An Advisory Panel was created in 1996 and their meetings were facilitated by the Gwinnett County Department of Public Utilities (DPU) with assistance from an environmental consulting firm. At a public meeting, attendees were polled and identified seven categories of stakeholder groups for the Advisory Panel:

- Homeowners associations
- Business community
- Development interests
- Large water users
- Gwinnett County cities
- Environmental organizations
- Citizens-at-large

[11] Matlock, Glenn, staff writer, December 3, 2000, *Gwinnett Daily Post*, untitled emailed article.
[12] Editorial staff writer, October 1, 1997, "Gwinnett plant to show discharge not all bad," *The Atlanta Journal and Constitution*, pgs. 12A.
[13] Soto, Lucy, February 22, 1996, "Water and Waste; Downstream Currents: As metro Atlanta grows, the rest of the state is taking a 'Don't flush on me' attitude." *The Atlanta Journal and Constitution*, pg. 04B.
[14] Shelton, Stacy, June 3, 1999, "State holds off on sewage discharge OK," *The Atlanta Journal and Constitution*, pg. 1JJ.
[15] Interview with Lake Lanier Association official, July 2001.
[16] 1998, *Gwinnett County, Georgia, Water and Wastewater Master Plan Update; Executive Summary*, Camp Dresser & McKee, pg. ES-1.

Representatives were "to act as a spokesperson for the group, representing the group's interest. The representatives were responsible for conveying information from meetings to other interested members of the stakeholder group."[17] Meetings were held the first Tuesday of each month, in the evening, for 18-months. The meetings were open to the public, but primarily sought input from panel members, although representatives from the Gwinnett Council for Quality Growth and the Chattahoochee Riverkeeper regularly attended. Dissenting opinions on any of the topics discussed were preserved for inclusion in the master plan update report.

The goals and objectives developed by the Advisory Panel included:

- Improve reliability of water and sewer system
- Develop strong maintenance and rehabilitation programs
- Protect public health and the environment
- Plan for water and sewer capacity proactively
- Minimize the negative impact of new facilities on neighborhoods and the environment
- Develop alternate water sources
- Pursue regional opportunities
- Manage water and wastewater demand
- Provide a high level of service at an optimum cost

The Advisory Panel evaluated all water supply and wastewater alternatives to indicate their preference in the near-term (present – 2010), mid-term (2010 – 2030), and long-term (2030 – 2050). The year 2010 was viewed as an important point in time, since water supply sources had not been firmly established to meet increasing demands after 2010. When considering water supply, the Advisory Panel expressed a:

- Preference for the continued use of Lake Lanier as a water supply source in the near, mid, and long term.
- Blended reuse was considered a secondary alternative in the long term.

The wastewater preferences were generally more disperse among the Advisory Panel members than their preferences for water supply. Nonetheless, the Advisory Panel made recommendations regarding wastewater:

- Given the quality of treated wastewater effluent from the NAWRF, non-potable reuse be "pursued vigorously" through all time periods.[18]
- Continue to seek conversions from septic tanks to public wastewater treatment.
- Discharge into the Chattahoochee River in the near-term was preferred, with a second option being discharge into Lake Lanier.
- Increased preferences for blended reuse in reservoirs for the mid- and long-term planning horizons.

---

[17] 1998, *Gwinnett County, Georgia, Water and Wastewater Master Plan Update; Section 3: Citizen Participation*, Camp Dresser & McKee, pg. 3-2.
[18] 1998, *Gwinnett County, Georgia, Water and Wastewater Master Plan Update; Executive Summary*, Camp Dresser & McKee, pg. ES-4.

The Advisory Panel members actively wrote and edited the master plan update document.

During the master plan update process, county officials facilitated the public meetings. A county official noted that it made it difficult to "maintain balance among the participants and prevent some individuals from dominating" and to remain "neutral" as the facilitator and the county. However, at the same time, the official felt it was important to "not put hired help between us and the citizens…. If we wanted to build trust, [obtain] a higher degree of credibility, then we could not hide [behind] a professional facilitator."[19]

CITIZEN ADVISORY BOARD OVERSEEING DISCHARGE

The County took a "partnering" approach with citizens on the project.[20] Formed in April 1996 and continuing today, the Citizen Advisory Board has oversight responsibilities at the NAWRF, with an emphasis on proper operations and meeting the effluent standards. While there were a few common members with the master planning process Advisory Panel, the Citizen Advisory Board is an independent group with a distinct role. It serves as a communication channel between the public and the utility. The Citizen Advisory Board controls its own $50,000/year budget. The Citizen Advisory Board has spent the funds on sampling, technical review of plans and designs, and other oversight activities. The Board currently has a subcommittee on discharge monitoring.

The Citizen Advisory Board's purpose is to facilitate communication process between the public and the county about the plant and to oversee discharge monitoring program. Furthermore, their role is specific, as one member said, we "focus on the plant...not where [the discharge] goes."[21] For example, the Board was "instrumental in having the County Board of Commission add a county resolution to have annual budgeting [for the plant] include percentage increase labor costs for retraining...then budget accountants cannot strike training from annual budgets." Retraining is seen as important to address long-term maintenance at the facility; as one Board member put it, we "have a highly qualified group running the plant, but if they don't continue training... they could become complacent."[22]

A county public utilities official noted that he wanted individuals to stay involved for at least five years, until the NAWRF plant was completed.[23] While the Board's members are elected to 2-year terms, there has been little interest from new members and old members are remaining on for longer terms. Recently, two new alternates were added from other counties bordering on Lake Lanier.

NAWRF EXPANSION

Given the expanded capacity needs of a growing Gwinnett county, the needs of neighboring counties, and the state's interest in a regional approach to the problem, Gwinnett County pursued a 40 mgd expansion of NAWRF prior to completing the 20 mgd facility. The master planning process encouraged the expansion. However, rather than discharging into the Chattahoochee, the

[19] Interview with a county official, May 2001.
[20] Interview with county official. May, 2001.
[21] Interview with Citizen Advisory Board member, July 2001.
[22] Ibid.
[23] Matlock, Glenn, staff writer, February 28, 1996, *Gwinnett Daily Post*, untitled emailed article.

permit requested discharge into Lake Lanier, through a 10-mile pipeline. The state EPD directed the discharge into the Lake, consistent with their water resources management policy to discourage cross-basin transfers and the interstate compact being negotiated with Florida and Alabama. County officials believe "it simply makes good water resources sense for the region to sustain flows below Lake Lanier for us to return reclaimed water to the Lake."[24] A journalist confirmed that the idea of retaining as much waste in the county as possible before it flows downstream is an "argument [that] clicks with people."[25] Lake Lanier is also the drinking water supply for more than 2 million people in the metro Atlanta region.

Supporters included staff writers with *The Atlanta Journal and Constitution* who called Gwinnett the "'good guy,'...something of a poster child, for good management of its water resources... the guys in the white hats in the water wars."[26] In 1999, the state EPD initially delayed action on the permit request until after setting water quality standards for Lake Lanier, specifically for chlorophyll A, nitrogen, phosphorus, dissolved oxygen, and fecal coliform bacteria. By early 2000, the standards were released (zero sum increase in phosphorus) and by July, 2000 the state issued a permit for NAWRF's discharge into Lake Lanier. In fact, county officials felt they had convinced the originally-skeptical EPD officials that the NAWRF could meet a higher standard and "raise the bar for Georgia."[27]

OPPOSITION TO NAWRF DISCHARGE INTO LAKE LANIER

Civic and environmental groups had hoped for stronger water quality standards.[28] The Lake Lanier Association, representing businesses, homeowners, and recreational users, expressed concern, "It's not just a chance; it will happen eventually that stuff that doesn't belong in the lake will end up there. The disadvantages outweigh the advantages."[29] "It's the first of probably 300 million gallons [in discharge requests] over the next 30 years," claimed a Lake Lanier Association representative.[30] In December 2000, the Lake Lanier Association, local Sierra Club, Headwaters Conservation, and the Upper Chattahoochee Riverkeepers filed an appeal of the state permit. They are challenging the permit on the water quality standards and a design change in the location of the underwater discharge point in the Lake. They are concerned that the discharge will further degrade the lake, create algae blooms, and increase the water temperatures to unhealthy levels for fish.[31] An appellant claimed, "we are not concerned about the plant... just about the reservoir."[32]

---

[24] Email survey response from county official, July 2001.
[25] Interview with journalist, May 2001
[26] Brack, Elliott, August 15, 1999, "Returning treated water to lake is best use of resources." *The Atlanta Journal and Constitution*, pg. 4JJ.
[27] Interview with county official, May 2001.
[28] Seabrook, Charles, January 27, 2000, "State sets Lanier water quality standards," *The Atlanta Journal Constitution*, pg. 4B.
[29] Cory, Joe, staff writer, July 14, 1999, *Gwinnett Daily Post*, untitled emailed article.
[30] Shelton, Stacy, December 3, 1999, "Agency delays county's bid to expand sewage discharge," *The Atlanta Journal and Constitution,* pg. 1JJ.
[31] Shelton, Stacy, February 15, 2000 "County moves ahead on sewer project," *The Atlanta Journal and Constitution,* pg. 2JJ.
[32] Interview with a Citizen Advisory Board member, July 2001.

Representatives of the Lake Lanier Association have expressed frustration over communication with the County; "County won't tell use anything... no specific answers [to our questions] and no open communication."[33] Meanwhile, the County have viewed the permit appeal as limiting communication, "since the LLA is an appellant to the Lake discharge permit, communication directly with the LLA is not occurring."[34]

Opposition has not arisen on the Citizen Advisory Board. As one member explains, "the Citizen Advisory Board doesn't have anything to do with water going to the lake. That's more of an environmental concern and the Citizen Advisory Board looks at the plan...just deals with issues related to the plant."[35]

NAWRF BECOMES THE F. WAYNE HILL WATER RESOURCE CENTER

Also in December 2000, the 20 mgd NAWRF was dedicated and renamed the F. Wayne Hill Water Resource Center, after the Gwinnett County Commission Chairman, the "driving force behind the project." Chairman Hill noted, "I have strived to improve the environment – to improve the quality of wastewater going back into the river. And I believe this plant sets the standard for Georgia." The Governor of Georgia requested to attend the dedication ceremony and stated his support for the plant expansion.[36]

MEDIA COVERAGE

A member of the Citizen Advisory Board claimed that the media "reports information rather than editorializes" and that the "local paper provided very good coverage from time to time when issues of significant interest [arose]. But in general, the media has "not found anything sensational to report."[37]

County officials stated that they tried to tell everyone when mistakes were made, not try to hide anything but lay it all out there. "We put the bad news out in front of people… and it buys press credibility and with our oversight board."[38] A journalist confirmed, "[the government official] is adept about describing technology and he's a straight forward guy, who will be straight with you about any potential problems."[39]

PROJECT IMPLEMENTATION

The County took a partnering approach with contractors on the project. Contractor teams were assembled at each stage of the process (i.e., design, excavation, construction) and reached a partnering agreement. They established performance criteria and agreed that "when mistakes were made, there would be no finger pointing, no litigation, but rather, we'd fix it."[40] For example, when it was determined that more dirt had to be moved than originally estimated and

---

[33] Interview with Lake Lanier Association official, July 2001.
[34] Email survey response from county official, July 2001.
[35] Interview with Citizen Advisory Board member, July 2001.
[36] Matlock, Glenn, staff writer, December 7, 2000, *Gwinnett Daily Post*, untitled emailed article.
[37] Interview with Citizen Advisory Board member, July 2001.
[38] Ibid.
[39] Interview with local journalist, May 2001.
[40] Interview with county official. May, 2001.

when contractor bids came in higher than anticipated, the county "let everyone know, put the word out. Then we said, `we are going to overcome this.'"[41]

The plant has experienced three spills during start-up operations when monitoring sensors failed. The first in December, 2000 involved 5,000 gallons of treated wastewater and two more in March 2001 occurred when a solids storage tank overflowed. On March 7th, 4,000 gallons of spilled materials were collected in retention areas, with about 1,000 gallons reaching Ivy Creek a half-mile away.[42] Later in March a third spill of 2,400 gallons occurred when a valve malfunctioned and in May, 2001 a fire after a gas leak caused minor burns for two plant workers.[43] A county official noted, "the overflows at the plant were all start-up problems. The plan has consistently done better than its permit limits."[44]

Nonetheless, overall it has been difficult for the Citizen Advisory Board, the Lake Lanier Association and the County to generate much interest in the general public. A Citizen Advisory Board member noted that there is "general public apathy... [and] as long as someone else is watching it, they just as soon be home with a beer in front of the TV."[45] A Lake Lanier Association representative stated that the facility is "not in the public's attention."[46] A county official counters that there is public interest in the county.[47]

---

[41] Ibid.

[42] Ibid. and Matlock, Glenn, staff writer, March 8, 2001. *Gwinnett Daily Post*, untitled, emailed article.

[43] Warren, Bet, May 19, 2001, "Flash fire at sewer plant injures two," *The Atlanta Journal and Constitution*. Nurse, Doug, May 31, 2001, "$4,800 fine to be paid for sewage plant spill; Two other accidents may draw penalties." *The Atlanta Journal and Constitution.*

[44] Comments from county official. August 2002.

[45] Interview with Citizen Advisory Board member, July 2001.

[46] Interview with Lake Lanier Association representative, July 2001.

[47] Comments from county official, August 2002.

CONCLUSION – CURRENT STATUS

While the permit for the expansion of the F. Wayne Hill Water Resource Center remains under appeal, the county has issued contracts for design and has moved ahead with the expansion planning and construction, aiming for start-up in 2005. There is interest in the state and with other counties to use the facility as a regional plant. The permit appeal is moving through the Administrative Law offices of state government, with a hearing scheduled for the fall.

TIMELINE:
GWINNETT COUNTY, GEORGIA - F. WAYNE HILL WATER RESOURCE CENTER

1990:
-Residents sue County over wastewater treatment plant.
-500,000 gallon lightning induced sewage spill.

1995:
-Site of new wastewater treatment plant rejected.
-County approves building of NAWRF.

1996

Aug '96:
State EPD issues draft permit.

April '96:
Citizen Advisory Board formed for NAWRF.

Late '96:
County begins updating Water and Wastewater Master Plan.
-Advisory Panel formed to provide citizen input into Master Plan.

1998

June '98:
-Master Plan revisions complete.
-Advisory Panel concluded.

1998:
NAWRF construction begins.

1999

April '99:
NAWRF expansion proposed.

Late '99:
State EPD delays action on permit to set water quality standards for lake.

2000

Jan '00:
Water quality standards released.

July 00:
Permit issued.

2001

Dec '00:
-Civic and environmental groups file appeal of the state permit.
-NAWRF dedicated and renamed.
-5000 gallon spill at plant

March '01:
Two spills at plant.
May '01:
-County pays fine for spill.
-Small fire at plant. results in minor injuries to two workers.

Currently:
-Permit appeal hearing in Fall
-Moving ahead with expansion planning.
-Design contracts issued.
-2005 start-up planned.
-Citizen Advisory Board continues meeting.

III. CITY OF SAN ANTONIO, TEXAS: WATER RECYCLING PROGRAM

Abstract:

The San Antonio Water System (SAWS) relies on the Edwards Aquifer to meet the drinking water needs of the San Antonio metropolitan area. In the recent past, droughts and rapid population and economic growth strained the aquifer, causing it to be drawn down faster than it could be replenished by scarce rainfall. In 1996, SAWS Water Recycling Program was launched in part to educate communities about the need for a water supply to supplement the Edwards Aquifer. The centerpiece of the program was a $125M pipeline infrastructure to deliver approximately 35,000 acre-feet/year of recycled water from the city's existing water reclamation plants to commercial, industrial and agricultural customers. Use of recycled water is predicted to reduce demand on the Edwards Aquifer by 20 percent, enough to satisfy the annual water needs of up to 80,000 households. While some communities initially perceived the program as a step toward direct reuse, most accepted the benefit of reduced demand on the Edwards Aquifer. The SAWS Community Outreach Plan sought to keep stakeholders involved and informed during each phase of the program, including concept development, environmental review, design, and construction.

INTRODUCTION: WHY WATER RECYCLING IS AN ISSUE IN SAN ANTONIO

The City of San Antonio, like many fast growing communities in the southern central plains, regularly experiences significant water use constraints. In fact, the demand for water far exceeds the historic average replenishment rate for the city's sole water source, the Edwards Aquifer. Furthermore, San Antonio's population continues to grow.

In 1996, an extended drought triggered mandatory rationing under an aquifer management plan, including surcharges for residential water usage greater than 17,205 gallons/month. The city council terminated the surcharge four months later, after "a good deal of community unrest and confusion."[48] Many citizens simply did not believe a water shortage existed.[49] On two occasions in the 1990s, plans for a new reservoir were voted down.[50]

Furthermore, groundwater pumping for commercial, industrial and irrigation purposes reduced flow to natural springs. Spring flow fed several federally protected endangered species habitats as well as Salado Creek, which flows through the city's River Walk area, a major tourist attraction. The Sierra Club filed a federal lawsuit seeking to impose strict groundwater pumping limits on San Antonio and other high-volume groundwater users to restore the spring flow. The lawsuit compelled the city to develop an aquifer management plan to address environmental concerns. In addition, state legislation passed in 1997 mandating comprehensive water resource planning. (see timeline on page 20)

---

[48] March 4, 1997, *Water Recycling Program: Community Outreach Plan*, Prepared by Katz & Associates, pg. 2.
[49] Ibid. pg.11
[50] Interview with SAWS official, May 2001.

How San Antonio Responded to Its Water Constraints

In June 1996, while instituting unpopular water rationing measures, the city council appointed a "broad and diverse community-based group of 34 representatives" to develop a water policy approach for San Antonio.[51] The Citizens Committee on Water Policy presented their final report in January 1997, outlining a comprehensive framework that emphasized the importance of continued citizen involvement in the evaluation of current and future options for water resources. The report recommended several immediate steps, including:

- water conservation;
- surface water rights study;
- water recycling;
- specific responses to the low flow at the natural springs;
- irrigation suspension;
- weather modification;
- aquifer optimization;
- desalination water study;
- water quality protection measures; and
- regional cooperation.

In addition, the citizens committee established clear criteria for the process of considering current and future water resource options. Those criteria included:

- trust and credibility of process;
- community and unity promotion;
- availability and reliability of water supply;
- economic opportunity;
- environmental benefits and sustainability;
- efficiency and affordability of outcomes;
- governance and control of options; and
- fairness and equity in process and outcomes.

SAWS Water Recycling Program

In response to the citizens committee's recommendations and criteria and the broader water context in San Antonio, SAWS developed the Water Recycling Program. The goal of the SAWS Water Recycling Program was to increase the availability of water from the Edwards Aquifer by supplying recycled water to high-volume commercial, industrial and agricultural users. The recycled water would also augment flows in the San Antonio River and Salado Creek. The program would deliver 35,000 acre-feet of recycled water/year, replacing up to 20 percent of existing demand for aquifer water, enough to satisfy the annual needs of up to 80,000 families. Historically in San Antonio, recycled water has been used for approximately 30 years in the cooling-tower operations of the city's power plants.

---

[51] March 4, 1997, *Water Recycling Program: Community Outreach Plan*, Prepared by Katz & Associates. pg. 4.

The supply of recycled water comes from four existing SAWS water recycling facilities. At these facilities, wastewater goes through primary and secondary treatment, and an advanced process that includes chemical-assisted filtration, disinfection, and dechlorination. No additional costs are associated with building the plants to recycle water; rather, the expense is for constructing a separate set of recycled water pipelines. The one-time capital cost to build the delivery system is approximately $125 million. Operation and maintenance costs are estimated to equal current costs of delivering Edwards Aquifer water through the drinking water system.

Significant demand for the recycled water existed during the planning stages of the water recycling program, as commercial, industrial, and agricultural users saw a benefit to having a guaranteed water supply even during drought conditions. Customers included:

- City of San Antonio;
- Brooks and Lackland Air Force Bases;
- Fort Sam Houston;
- Witte Museum;
- San Antonio National Cemetery;
- Trinity University;
- Greater Kelly Development Corporation;
- USAA; and
- Several golf courses.

Phase I plans and designs were approved in October 1996, including two pipeline branches, pump stations, and a system to control operations.

WATER RECYCLING PROGRAM'S COMMUNITY OUTREACH PLAN

SAWS's Community Outreach Plan is a component of the Water Recycling Plan. The goal of the outreach plan, developed in early 1997 was to "increase public education about recycled water and to build broad-based community support for the program as a viable and cost-efficient supplemental water resource which benefits the entire San Antonio community."[52] The plan sought to build and maintain a partnership with the community to ensure effective project implementation and long-term success. It stated:

> Building and nurturing a partnership with the community is an ongoing and constantly changing endeavor. It is important the agency is committed to building the public's trust by designing a public participation process that is fair, allows for two-way communication, solicits honest community feedback, and is willing to incorporate public input into the final program design. (SAWS Community Outreach Plan, pg. 1)

The plan considered the critical factors for project success to be customer marketing, technical feasibility, and public acceptability. The recycled water would be made available to commercial, industrial, and agricultural customers rather than to residential users. A strong customer base

[52] Ibid. pg. 8.

was viewed as important to the success of the program. Technical feasibility referred to the cost-effectiveness and engineering feasibility of the system, which were viewed as critical elements to the program's success.

Public acceptance and support was characterized by the community outreach plan as "perhaps … the most important factor in ensuring the project's timely completion and overall success. As a result, the development of the program includes public input at every key decision-making point and demonstrate an overall sensitivity to public issues and concerns."[53] Preliminary public workshops, focus groups, public opinion surveys, and stakeholder interviews were conducted in late 1996 and into 1997. Findings included:

- Majority of respondents felt there was not a water shortage, but that water conservation and water recycling were good methods of protecting and preserving San Antonio's current water supply.
- There was concern about the project cost and who actually benefited from the program.
- Residents felt there was a lack of reliable information sources on water.
- Residents believed strongly in protecting the aquifer recharge zone, recycling water and storing more in surface reservoirs, and assuring a dependable water supply.
- One in five (20.1%) of a surveyed sample (N=600) were aware that water recycling involved the "reuse of water for nonpotable purposes."
- Over half, 52.4%, agreed strongly that water recycling was a way to reduce shortages and save aquifer water for drinking water.

The consulting firm conducting the research concluded that SAWS needed to reinforce the value and benefits of water recycling and provide more information about the specific components of the SAWS program.

Project Implementation

The value and benefits to all of San Antonio have been repeated in newspaper coverage of the water recycling program since 1997. For example:

> 1997: Quoting a SAWS official: "Just remember that with every drop of recycled water we use, we save aquifer water… It's not really for the residential customer. But this program provides a cost benefit to everyone."[54]
>
> 1998: Using recycled water for irrigation, cooling-tower operations, manufacturing and other non-drinking purposes will free more than 11 billion gallons of water a year that come directly from the Edwards Aquifer… That would supply the annual water needs of up to 80,000 families.[55]
>
> 1999: The recycled water will replace the drinking-quality Edwards Aquifer water now used by the [commercial and industrial] customers. The "saved" drinking water, amounting to about 20 percent

[53] Ibid. pg. 9.

[54] Cisneros-Lunsford, Anastasia, July 16, 1997, "Water recycling project to trim industrial, commercial usage," *San Antonio Express-News*, Sun – Southside pg. 1

[55] Barrios, Joseph, May 13, 1998, "Traffic flow v. water flow: Pipeline construction to close roads but long-term watering benefits," *San Antonio Express-News*, Metro, pg. 1.

> of the utility's annual use, will be available for other customers and for growth... [Commercial and industrial] customers pay the same rates as they would for Edwards Aquifer water, but have an incentive to use recycled water where possible because the recycled water, unlike Edwards water, is not subject to cutbacks in time of drought.[56]
>
> 2001: The system is expected to free enough Edwards drinking water for up to 80,000 families a year... The dollar rate [for the recycled water] is comparable to the rate for potable water...[57]

Though the water recycling program received general support from community groups across the city, some lower-income residents where opposed to plans for laying pipelines through their community. A neighborhood association member noted in 1998, "I know some people have expressed concerns that this is another way to get people to drink recycled water. Since the southside has traditionally been the place where everything gets dumped whether you like it or not, I think people... still have their suspicions."[58] In fact, in 1999 southside residents and political leaders opposed a SAWS plan to apply biosolids[59] to 1,500 southside acres.[60]

SAWS managers did consider requests from some residential customers for access to the recycled water. These requests came from residents of an affluent area that the pipeline would pass through. The residents reasoned that if their streets were going to be disrupted to lay the main pipe, access pipes ought to be laid as well to provide residents a source of lawn water that would not be subject to drought restrictions. Upon consideration, SAWS managers denied the requests and remained committed to their plan to serve only commercial and public properties. SAWS managers believed it was not fair to provide access to this community, without providing a similar service to all residential communities along the two pipelines.[61]

---

[56] Needman, Jerry, December 8, 1999, "Gray water tops estimates: Latest cost put SAWS bill at $125.3 million." *San Antonio Express-News*, Metro, – South Texas. pg. 7B.
[57] Prendez, Linda, February 14, 2001, "Recycled water available; Pipes being laid in Leon Valley," *San Antonio Express-News*, Metro – Sun Northwest, pg. 1H.
[58] Barrios, Joseph, May 13, 1998, "Traffic flow v. water flow: Pipeline construction to close roads but long-term watering benefits," *San Antonio Express-News*, Metro, pg. 1.
[59] Biosolids are the nutrient-rich organic materials resulting from the treatment of sewage sludge (the name for the solid, semisolid or liquid untreated residue generated during the treatment of domestic sewage in a treatment facility). Source: U.S. Environmental Protection Agency, Office of Wastewater Management web page, http://www.epa.gov/owm/bio.htm.
[60] Barrios, Joseph, June 11, 1999, "South Side a boil at disposal plan," *San Antonio Express-News*, Metro, pg. 1.
[61] Interview with a SAWS official, May 2001.

CONCLUSION – CURRENT STATUS & OVERVIEW

The program broke ground in September of 1997 and installed the first recycled water pipe the following month. Water began to flow to the first recycled water customer, City of San Antonio Parks and Recreation, in January 1999. By April 1999, Phase I was 50% complete. Delivery to the program's first private customer, Trinity University, began in the summer of 2000. Currently, the recycled water pipeline is more than 80% complete. Construction of Phase II has begun to connect the two sections of the system.

TIMELINE:
CITY OF SAN ANTONIO, TEXAS WATER RECYCLING PROGRAM

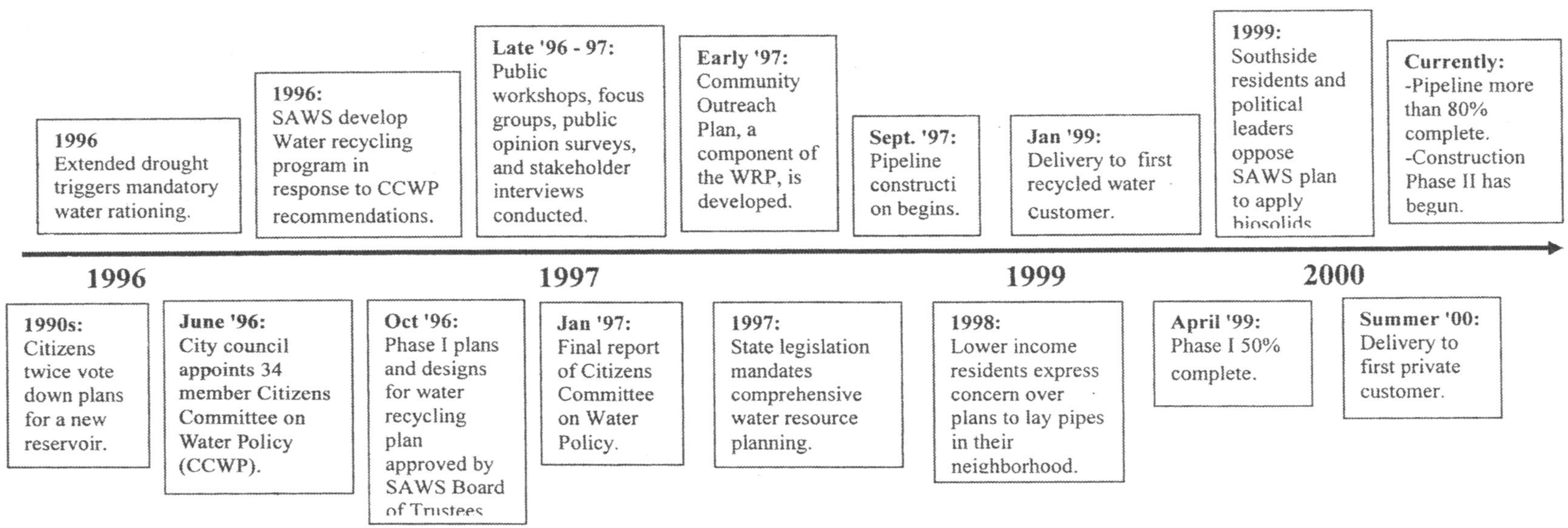

IV. CITY OF SAN DIEGO, CALIFORNIA: WATER REPURIFICATION PROJECT

Abstract:

In 1990, ninety percent of San Diego's water supply was imported. The volume imported was expected to remain the same or decrease as commercial and residential demands for water were expected to increase. In 1994, a legal settlement required the city to build a water reclamation system with a 45 million gallon per day capacity. Limited demand for the reclaimed water for non-potable uses, however, would result in millions of gallons being dumped into the ocean. To address this concern and also reduce reliance on imported water, the city and the county water authority proposed a project that would infuse reclaimed water into the potable-water supply. The proposal moved forward for several years, gaining approval from panels of technical experts and support from the California Department of Health Services. By 1999, however, political and public opposition stopped the proposed project. Concerns were raised about racial and economic bias associated with the project design and about public health and project costs. Today there remains considerable disagreement between parties over how and why events played out the way they did.

Introduction: Why Water Repurification is an Issue in San Diego

Supply and Demand

For several years prior to the 1990s, Southern California had benefited from surplus water in the Colorado River Aqueduct. This imported water made up ninety percent of San Diego's supply, though polls showed that most of the public did not realize this fact. However, with cities upstream in Arizona and Nevada continuing to grow and use more water, Southern California had to find a way to live within its water allocation. By 1990, while its supply from imported sources was expected to remain the same or decrease, San Diego's demand for water was expected to continue to grow.

Pilot Project

Recognizing the likelihood of future constraints on the water supply and the limitations of relying so heavily on imported water, San Diego initiated a pilot project to test the viability of a potable water reuse system in the early 1980s. The system operated from 1984 to 1993. The water from the system was not distributed for potable use; rather, the project provided data on the mechanical reliability of the system and the safety of the reclaimed water. Different views are expressed by government officials, elected officials, technical experts, and others interviewed over the degree to which the pilot demonstrated success and was used in the public education and outreach efforts.

Court Settlement

The pilot project research analysis was being finalized in 1994 at the same time that San Diego was reaching settlement of a lawsuit with the U.S. Environmental Protection Agency. As part of the settlement, San Diego agreed to build a water reclamation system with a 45 million gallon per day capacity. Work was already underway on the North City Water Reclamation Plant (NCWRP), which was intended to supply water to customers for non-potable uses -- the NCWRP was completed in 1997. However, to distribute the reclaimed water, more than 65 miles of pipe would need to be laid beneath city streets at an estimated cost of $162 million. Together, the NCWRP and the distribution pipes were required to comply with the legal settlement with the U.S. EPA.

While the NCWRP would generate 45 million gallons per day, the city anticipated problems using that much reclaimed water. They estimated that during months of high water demand, most of the water would be used for irrigation. However, during cooler winter months, demand would be so low that most of the reclaimed water would be discharged to the sea. (see timeline on page 28)

How San Diego Responded to Its Water Constraints

To address the city's concerns about discharging excess reclaimed water to the sea, to reduce the city's reliance on imported water, and to guard against potential future water shortages, the city and the county water authority proposed a project that would infuse reclaimed water into the potable-water supply. The city and county saw this as an important element of a more comprehensive water resource management strategy. The Repurified Water Project, as it came to be called, would allow San Diego to reuse an estimated 45 percent more of the reclaimed water than would the project if it were only devoted to non-potable uses. The project would use treated water produced at the NCWRP and store the water for subsequent potable use at San Vicente Reservoir.

The city and the water authority presented a preliminary feasibility study to the California Department of Health Services (DHS) in August of 1993. DHS indicated that it was "cautiously optimistic" that the Repurified Water Project could be approved and requested a comprehensive feasibility study.[62] The water authority and the city agreed to work jointly to develop the study, with the water authority being the lead.

To assess community attitudes toward repurified water, project planners implemented a research program including telephone surveys, focus groups, and community leader interviews. Public education efforts were conducted including a media education campaign, establishing a review committee and developing informational materials and educational presentations.

State's Independent Advisory Panel

Shortly after the water authority selected a contractor, San Diego County Water Authority (SDCWA) selected nationally recognized experts in the fields of water treatment and public health to serve on an independent advisory panel (IAP). The IAP provided technical oversight of the proposed Repurified Water Project and acted as an advisor to SDCWA in its review of the project. The panel completed its review of the proposed project and submitted comments to SDCWA and DHS in the spring of 1994, recommending that the project continue and suggesting specific modifications.

[62] November 17, 1994. *Repurified Water Review Committee Final Report to the San Diego County Water Authority Staff*, pg. 21.

After incorporating the recommendations of the IAP, the water authority submitted its final proposal to DHS in June 1994. The proposed project would treat approximately 20 million gallons per day of reclaimed water to a level suitable for introduction into a raw water reservoir. The process proposed would remove harmful constituents from the water through a series of multiple barriers including:

- primary, secondary and tertiary treatment of wastewater at the North City Water Reclamation Plant (NCWRP);
- advanced water treatment using reverse osmosis membranes and disinfection with ozone/hydrogen peroxide (nutrients that may affect reservoir water quality would be removed through an ion exchange process);
- conveyance from the NCWRP through a 23-mile pipeline to the 90,000 acre-foot San Vicente Reservoir;
- retention in the reservoir for an average of two to three years;
- final filtration and disinfection at the Alvarado Water Treatment Plant; and
- a reliability assurance plan and monitoring program that would identify and divert any repurified water that did not meet standards.

Repurified Water Review Committee (RWRC)

After the SDCWA submitted its final proposal to DHS, the Repurified Water Review Committee (RWRC) was formed. The committee's mission was to assist water authority staff in examining all aspects of the draft proposal that were potentially of interest to San Diego residents. The committee consisted of 16 members and one special advisor, representing a broad cross-section of the San Diego community. Members represented:

- San Diego Council of Bass Fishing Clubs
- Industrial Environmental Association Water Committee
- San Diego Restaurant Association
- San Vincente Reservoir, Recreational users
- San Diego County Black Chamber of Commerce
- Greater San Diego Chamber of Commerce Government Affairs Steering Committee
- San Diego Taxpayers Association
- Sierra Club, Water Reclamation Committee Chair
- UCSD School of Medicine
- Kelco
- Citizens Coordinate for Century 3
- Special Assistant to San Diego City Councilmember on Senior Issues
- Biomedical Industries Council
- Metropolitan Wastewater Department, Oversight Committee
- Parent Teachers Association
- Asian Business Association

After examining the issues related to the proposal, they would provide input as to the suitability of repurified water as a supplement to the region's water supply.

RWRC members toured several of San Diego's water treatment and storage facilities. They also received briefings on:

- The water authority's water resources plan
- Population forecasts
- Water and wastewater overview
- Water repurification history and issues
- DHS comments
- Membrane technology and reverse osmosis
- Project costs and economics
- Community issues and public perception
- Health issues

Public awareness was gauged and public concerns and interests were identified through stakeholder interviews. Based on the information, the RWRC recommended that targeted community outreach should continue. The RWRC report indicated that:

> The [water reuse] subject matter should be concise and [documents should] use simple graphics. Special emphasis should be placed on describing the total water cycle in order to help the lay person understand the process. The Committee suggested that two messages should be reinforced: the current imported water supplies have been previously used, and San Diego's supply is 90 percent dependent on imported water, especially during a drought.[63]

RWRC held six meetings over the next five months and submitted its final report to the water authority in November. The committee concluded:

> There is sufficient information available to determine the suitability of water repurification as a supplement to the San Diego region's water supply. Additional planning, economic analysis and environmental studies should proceed.[64]

Following review of the RWRC report by the water authority's board of directors, the project entered the detailed planning phase. In 1995, as planning for the project continued, lead responsibility was transferred from the water authority to the City of San Diego's Metropolitan Wastewater Department. Project planners expected to move to the design phase in 1996, begin construction in 1998 and begin supplementing local drinking water supplies with repurified water in 2000.

[63] November 17, 1994. *Repurified Water Review Committee Final Report to the San Diego County Water Authority Staff*, pg. 15.
[64] Ibid. pg. 3.

Emerging Opposition

Project staff periodically reported the status of the project at public hearings before the city council, though a city council staff person stated they were not kept informed and were "brought in at the ninth inning."[65] The editorial board of the local newspaper, *The San Diego Union-Tribune*, expressed opposition to the proposed project. In April of 1994, the newspaper had published a cartoon of a dog drinking from a toilet and a man behind the dog saying "Move over..."[66] The newspaper later began to refer to the repurification project as "toilet to tap." The phrase was widely used by those opposed to the project and loathed by project proponents.

Concern about the project among the public grew in 1997 and 1998 as several elected officials voiced their opposition. One official commented, "...it doesn't seem intuitively rational to move to [potable reuse] as the use when we know there is an extraordinary high level of public resistance."[67] A state assembly member spoke against the project at a press conference, using a toilet hooked to a drinking tap by a hose as a prop. Project staff expressed dismay that the elected officials "brought [the issue] to the public eye [in a negative way]" and shifted the discussion from the public health and technical issues to the political arena.[68] Meanwhile, the elected officials felt they were "just asking questions" about the project and facilitating public discussion.[69]

Legislation was passed requiring a public hearing before the project could continue. The hearing was well attended; however, one interest group representative described it as "disgraceful" because at the meeting, elected state officials openly blamed and criticized state DHS staff in a less than civil manner.[70]

Environmental Justice

Concerns over race and economic class bias surfaced as the city council member from the district surrounding the San Vicente Reservoir voiced his opposition to the project. The San Vicente Reservoir was the only reservoir in San Diego deemed large enough to accommodate the proposed amount of reclaimed water, and was the lowest cost option. It was perceived that the reservoir served the southern section of the city, where the population is predominantly African American and of lower income than that of the northern section. The reservoir served areas of the city of all economic classes and racial make-ups. The opposition viewed the proposal as another step in dividing the city along economic lines – "those who could afford to drink bottled water and those who could not, those who drink champagne and those who drink sewage," recalled a resident.[71] African American residents expressed concern that they were being made to serve as "guinea pigs," citing cases which they believed demonstrated that disease experiments had been conducted in the past on unknowing communities of African Americans in other cities.

---

[65] Interview with a city council staff member, May 2001.
[66] April 5, 1994. *The San Diego Union-Tribune*, pg. B-6.
[67] May/June 1998. "San Diego's Repurified Water for Industrial Use," *Update: Water Reuse Association of California.*
[68] Interview with a project staff member, May 2001.
[69] Interview with an elected official, May 2001.
[70] Interview with interest group representative, April 2001.
[71] Ibid.

Others opposed to the project raised issues of cost, arguing that less expensive alternatives existed, including desalination or importing water from additional sources. At the same time, some argued that city residents would be willing to pay the costs of distributing the reclaimed water for non-potable uses.

Project staff prepared information materials and made public presentations to explain the technical approach and merits of the project. Their efforts, however, were not always effective. In the opinion of one technical expert on the project, while "the city was highly concerned about the charge of environmental racism…they unfortunately…did not deal with it…[and] thought that by showing it was not true (i.e., potable reclaimed water was going to be distributed to rich and poor neighborhoods alike) they would win out."[72]

The deep-seated concerns and values of those opposed to the project made communicating the technical information and safety precautions challenging. As the opposition became more vocal, project staff felt they themselves became more reactionary and less proactive, responding to the concerns of others and unable to explain their own perspective.

In March 1998, the city council delayed the release of the project's environmental impact statement so that further study could be included and another independent panel could review the proposed project.

National Research Council's Potable Water Reuse Report

That same month, the National Research Council (NRC) released a report, *Issues in Potable Water Reuse*, based on an evaluation of several existing reuse projects and the feasibility studies of Tampa and San Diego's projects. The NRC concluded that "reclaimed waste water can be used to supplement drinking-water sources, but only as a last resort and after a thorough health and safety evaluation."[73]

The independent panel convened to review the proposed project included seven nationally recognized scientists. The objectives of the "Blue Ribbon Panel on Water Repurification" was to:

- review the status of the proposed project and evaluate the city's technical findings;
- recommend appropriate strategies for achieving and maintaining public health, system reliability, and technical feasibility; and
- study the recently released NRC report and consider the issues raised as they pertained to the proposed San Diego project.

[72] Email survey response from Blue Ribbon Panel member, July 2001

[73] March 11, 1998. "Reclaimed Waste Water Can Augment Municipal Drinking-Water Supplies But Uncertainties Must Be Addressed," accessed May 11, 2001 at: www.nationalacademies.org. Also see, Balint, Kathryn, March 12, 1998, "Science panel calls S.D. plan viable, but only as 'last resort'" *Union Tribune*, San Diego: CA.

The panel's final report to the city council's Natural Resources and Culture Committee, on September 16, 1998, stated:

> It is the unanimous conclusion of the Blue Ribbon Panel of Experts that water repurification as proposed by the City will provide a safe and appropriate supplemental drinking water supply for the City of San Diego.[74]

At the September meeting, the Natural Resources and Culture Committee voted five to zero to accept the panel's report.

More than thirty citizens attended the meeting to voice their opposition to the project. One community resident stated, "You will never convince me that scientists are acquainted with every pathogen and have the resources to eliminate them."[75]

Mayor's & City Council's Action

In spite of the Natural Resources and Culture Committee's acceptance of the Blue Ribbon Panel's report supporting the project, the mayor never placed the environmental impact statement on the council's voting docket, and in March 1999, the council omitted the water repurification project from its list of sewer projects to be funded. The following month, a unanimous vote by the Natural Resources and Culture Committee officially ended the project and focused efforts instead on distributing the reclaimed water for industrial and landscaping uses.

CONCLUSION – CURRENT STATUS & OVERVIEW

Although the settlement with the U.S. EPA requiring San Diego to build a reclamation system did not require the city to reuse any of the reclaimed water, the federal grant received to help pay for the NCWRP did set reuse goals. By the end of 1999, two years after the NCWRP began operation, the city had met its first goal of reusing ten percent of the reclaimed water. Three percent went to commercial and industrial customers, and seven percent was used for testing and operation at the NCWRP and irrigating NCWRP grounds. The city continues to work toward its goal of reusing twenty-five percent of the reclaimed water by 2003. The potable reuse project is not being pursued.

---

[74] September 16, 1998. *Findings and Recommendations Related to the Scientific and Technical Review of the City of San Diego's Water Repurification Project*, pg. 2.

[75] September 16, 1998. "SD Water Project," *City News Service*.

TIMELINE:
CITY OF SAN DIEGO - WATER REPURIFICATION PROJECT

1984-'93:
Pilot project to test viability of a potable water reuse system.

1990:
San Diego imports 90% of its water supply.

Aug '93:
'Cautiously optimistic' feasibility study presented to DHS on repurified water reuse (incl. potable reuse).
Oct '93:
Community wide public education campaign started

1994

1994:
-Legal settlement requires city to build water reclamation system.
-Pilot project research analysis finalized.
-Work begins on nonpotable water reuse plant (NCWRP) underway.

Feb. '94:
Independent Advisory Panel (IAP) formed
April '94:
Newspaper publishes editorial cartoon in opposition of project.

May '94:
-IAP completes review of reuse project. Recommends that the project continues. And, proposes modifications.
June '94:
San Diego County Water Authority (SDCWA) submits final proposal to DHS with the changes.

June-Oct '94
-RWRC formed.
-RWRC concludes enough info available to continue with

1995

1995:
Lead responsibility transferred to City of San Diego' Municipal Wastewater Department.

April '96:
DHS affirms pilot study findings that technology can produce high quality drinking water.

1997

1997:
NCWRP completed.

1997-98:
-Elected officials voice opposition.
-Public expresses concern.

1998

1998:
Race and economic class bias concern raised by San Vicente area residents

March '98:
City council delays release of EIS until further study by independent panel.

Sept. '98:
IAP unanimously approves project.

Late '98:
-Natural Resources and Culture Committee accepts IAP report.
-Citizens attend meeting and voice opposition.
-Mayor omits EIS from councils voting docket

1999

March '99:
City Council does not fund project.

April '99:
NRCC unanimously votes to end project and focus on nonpotable reuse for industry and landscaping.

Currently:
-No potable reuse
-Continuing to work toward non-potable reuse of 25% by 2003.

APPENDIX C

PUBLIC PERCEPTION & PARTICIPATION IN WATER REUSE:

WHITE PAPERS

JULY 2001

Troy W. Hartley, Ph.D.

TABLE OF CONTENTS

Table of Contents i

I. **Maureen E. Austin** C-1
Public Perception and Participation in Water Reuse Initiatives: Perspectives and Recommendations

II. **Juliana Birkhoff** C-4
Community Conflict over Water Reuse: Perspectives from Conflict Analysis and Resolution

III. **Caron Chess** C-7
Success in Public Participation, Internal Organizational Factors, & Context: Important Considerations in Water Reuse

IV. **James Conklin** C-10
Public Participation and Perception in Water Reuse Initiatives

V. **Margot W. Garcia** C-13
Using "Blue-Ribbon" or Advisory Committee

VI. **Troy Hartley** C-16
Communication Embeds Information Into Our Views of the World and Shapes How We Perceive Water Reuse

VII. **Mary McDaniel** C-19
Comments on, "Public Perception and Participation in Water Reuse: Literature Summary"

VIII. **Michael McGuire and Edward Means** C-21
Promoting Potable Reuse in a Difficult Public/Political Environment

IX. **Jeffery Stone** C-23
Ensuring Public Confidence in Recycled Water Programs

X. **Chris J. Wiant** C-24
Public Perception and Participation in Water Reuse: Discussion Paper

XI. **Authors' Affiliations** C-26

I. PUBLIC PERCEPTION AND PARTICIPATION IN WATER REUSE INITIATIVES: PERSPECTIVES AND RECOMMENDATIONS

Maureen E. Austin

Natural resource professionals are often asked to anticipate potential problems, make long-term projections, and apply concepts and principles today that will address future needs or concerns. The public, by contrast, does not routinely engage in forecasting behavior when it comes to natural resource use. This mismatch between how professionals perceive a natural resource issue and how the public views it is cause for considerable frustration on the part of both parties.

Public service announcements of water shortages, of the need to conserve water, of water reuse initiatives all must overcome powerful public perceptions that are reinforced through daily experience. Repeated and frequent experiences have great power in shaping human thought. For the public, the simple act of turning on the tap daily and finding plenty of available water disputes messages of a water shortage and the need for conservation.

Images and experiences that are particularly strong evoke emotions, thus powerfully impacting thought and behavior. Stories of wastewater spills and overflows, though rarely the norm, have a profound effect upon public perception. The editorial "toilet to tap" cartoon seen in the San Diego case study is a strong negative image which calls to mind many of the fears the public has concerning water reuse -- fears of water contamination, sewage spills, and unseen toxins in the water supply. Images such as these have the power to impact public perception in a negative fashion.

However, we're far more complex than our daily encounters and perceptions. When faced with new or unfamiliar information, we try to fit that experience into the knowledge we already have. This tendency is part of our mental wiring; from early childhood we have been parceling sensory input into internal images and representations, in short, making sense of our surroundings. The mind is quick and awareness of its process is generally unconscious. Usually this works well, but it can also lead to jumping to conclusions too hastily based on minimal information. Citizens may be quick to perceive a water reuse initiative as yet another example of government excess before learning about the proposal, the reasoning behind it, or the need for it.

Following are some recommendations, based upon principles of environmental and cognitive psychology, which are offered as a guide to encourage meaningful interaction among resource professionals and the public.

A. Providing Information to the Public

<u>Less is more</u>.

When seeking to educate the public about natural resource issues, there is a tendency for professionals to provide as much information as possible. The material presented can be overwhelming to an audience to whom the concepts are new and unfamiliar. This can cause confusion and antagonism, which disrupts meaningful interaction and communication. Present information on water reuse and conservation initiatives in small installments, with each installation building upon the information presented previously.

Keep communication, messages, and public service announcements jargon-free.

Dialogue among resource professionals often uses a vocabulary common to the profession but uncommon in everyday public dialogue. When presenting information about water reuse initiatives, limiting the use of professional jargon and explaining concepts in simple terms improves the interactions among professionals and the public.

In sharing information, make use of contexts and formats familiar to the public.

Our view of the world is strongly influenced by our local surroundings and social interactions. We tend to pay more attention to exchanges occurring within familiar contexts. Furthermore, people do not want to appear foolish when learning new information or trying new behaviors (i.e. water conservation) and familiar formats and contexts help alleviate this concern. Sharing new, technical, or complex information with the public is enhanced when presented using local and familiar communication channels and social networks. This can be accomplished in several ways including sharing information in neighborhood newsletters and publications, via neighborhood meeting venues, or using stories or testimonials shared between neighborhood residents, colleagues, and other social acquaintances.

Make use of images, photos, diagrams, and graphics.

Human thought and perception relies a great deal upon images in processing information from the surroundings and learning new concepts and ideas. The popular expression "a picture is worth a thousand words" illustrates nicely our tendency to rely on images in perception and understanding. Visual references to water shortages (e.g. photographs of low reservoir levels) and using simple graphics (e.g., City of San Diego case study) can help reinforce messages from professionals stressing the need to conserve water.

B. Asking Information from the Public

Seek public input in community settings.

Public hearings and meetings organized by professionals for the purpose of obtaining citizen input tend to stifle rather than support information gathering. The structure of the meetings is often very formal, beginning with presentation of the issues by professionals and followed by a period for public input. If the issue at hand is particularly complex (and most natural resource issues are), citizens can experience difficulty and frustration that can overshadow prime opportunities for sharing their insights and preferences. By making use of community settings where citizens already gather for neighborhood meetings and social functions, professionals may be less inclined toward formal presentation and can focus on dialogue and communication with citizens about their concerns. The input gathered in community settings allows participants to feel informed and a part of the process without being inconvenienced, and results in program guidelines and policies more closely aligned with public interests.

Seek public input using informal formats.

Seeking public input via surveys, focus groups, and dialogue about resource management plans allows individuals to state their preferences without having to fully understand the complexities of the resource issues. Giving the public options to choose from and rate according to their preference provides professionals with public input, encourages meaningful public involvement, and discourages the information overload that often results when well-meaning professionals overwhelm the public with too much information.

Kaplan, S. and Kaplan, R. (1982) *Cognition and environment: Functioning in an uncertain world.* New York: Praeger. (Republished by Ann Arbor, MI: Ulrich's, 1989).

Kaplan, S., Kaplan, R., and Ryan, R.L. (1998) *With people in mind: Design and management of everyday nature.* Washington, D.C.: Island Press.

II. COMMUNITY CONFLICT OVER WATER REUSE: PERSPECTIVES FROM CONFLICT ANALYSIS AND RESOLUTION

Juliana Birkhoff

The case studies in the Public Perception and Participation in Water Reuse Initiatives Project follow predictable patterns and exhibit classic conflict dynamics. These conflicts can also be managed better following the principles and practices of conflict resolution. In this concept paper I will briefly outline some of the learning's from conflict analysis and resolution that may help us understand these conflicts and thus, manage them better. By "better" management, I do not wish to imply that these conflicts should not occur. Social conflicts carry out critical social functions. Social conflicts define and express important debates about justice, aesthetics, health, politics, and economics. Furthermore, social conflict is useful for society. It is one of the ways that social change occurs. The existence of conflict is a sign of a healthy, diverse, society communicating different values about human community, choices about economics and politics, and visions for the future. However, conflicts can be managed to encourage better decisions and discourage escalation, stalemates, and violence. Unfortunately, individuals, groups, and institutions frequently fail to wage these struggles productively, to achieve their own interests, or to find outcomes that reach beyond the stalemate at hand (Simmel 1955, Coser 1956, Deutsch 1973, Collins 1975).

Conflict Beginnings

Troy Hartley's Literature Summary indicates that future conflicts over the reuse of water in American communities are likely to increase from economic and population growth. Conflict analysis and resolution research bolsters and helps us explain this prediction. Public policies and decisions over how a community should provide safe drinking water and water for other uses are likely to cause real controversy and conflict. Challenges that 1) differentially affect different members of the community, 2) touch upon important aspects of individuals' lives, and 3) upon which community members feel action can be taken are likely to cause conflict (Coleman 1957, Wehr 1979, Felstiner, Abel and Sarat 1981, Kriesberg 1982). For example, communities are not likely to engage in conflict over a flood because everyone is affected and because useful actions occur after the event. However, the Gwinett County, San Antonio, and San Diego case studies illustrate water reuse can differentially affect different community members, relates to critical community concern, health and can be affected by community members actions.

The Georgia, Texas and California case studies also exhibit another finding from conflict analysis and resolution theory and research; conflicts initiated by internal community issues are likely to follow previous community cleavages. These community divisions usually affect three areas of community life. The first of these is economic. Most American communities are deeply divided by economic issues. The second area, power and authority, reflects the disenfranchisement of many residents and the non-responsive political and bureaucratic structures in most American cities. Finally, conflicts proceed along previous divisions over cultural values or beliefs. Different values and beliefs over how decisions should be made, over risk, heath and danger, and aesthetics profoundly affect the course of these conflicts (Gaventa 1980, Douglas and Wildavsky 1982, Hirschman 1970, Ross 1993).

Similarly, conflicts also follow previous social divisions. Community members may respond to community issues and controversies along racial, class, or other social group cleavages. Each dispute over a community issue or decision replicates previous debates and fights and represents an ongoing battle between the different groups (Coleman 1957, Himes 1980).

Conflict Dynamics

Although conflicts begin many different ways and over a variety of issues, conflict analysis and resolution research demonstrates that conflict follows a predictable path and similar dynamics. Without intervention, conflicts proceed through of cycle of escalation that changes the issues and changes the social organization of the community. Typically, a specific issue such as whether or not a community should reuse water, can give way to more general issues about water use and policy as well as new and different issues including concerns over race and class bias and different land uses (Simmel 1955, Coleman 1957 Deutsch 1973).

Furthermore, without productive engagement disagreements over policy choices may escalate to antagonism between parties. These hostile interpersonal dynamics develop through psychological and social psychological processes. They may begin instrumentally, as one party seeks to differentiate itself and its view on the issues from the other parties. For example, in San Diego elected officials portrayed their actions as facilitating public discussion while project staff said it their behaviour represented political interests.

The community also goes through a predictable cycle as the conflict escalates. Opposing side may become increasingly polarized. New organizations and leaders may spring up to further different interests. Existing community organizations are drawn into the conflict and may increase polarized relationships. Finally, word of mouth communication increases as people begin to mistrust traditional sources of information and news.

Conflict Management

Conflict analysis and resolution theory and research points to strategies to productively manage community conflicts over water reuse. The case studies in San Antonio, TX and Gwinett County, GA exhibit some of these strategies. Both communities recognized that better decisions emerge when diverse interests, knowledge, and expertise are involved in public decisionmaking. However, all three case studies exhibit shortcomings in fully involving stakeholders in framing, analyzing, generating, and implementing solutions to complex public problems. If all those who are affected are not meaningfully involved, they will seek other ways of articulating and meeting their interests.

In each of the three cases, several other strategies could have improved outcomes. Conflicts are more productively handled by focusing on the substance of the problem separately from the problems between individuals. Everyone who is affected should also follow the complete steps of a conflict resolution process. This includes completely identifying and analyzing the problems before trying to solve them, identifying criteria for a good solution before developing solutions, assess and chose an option with all parties input and use the jointly derived criteria to implement and assess the solution.

Conclusion

This brief concept paper has outlined some of learnings from the field of conflict analysis and resolution that helps us understand the emergence, dynamics, and handling of conflicts over water reuse. I look forward to feedback from symposium members and advisory group members over how to incorporate these ideas into the framework and ultimately, better ways of managing water reuse decision-making.

References

Coleman, James Community Conflict, New York, NY: The Free Press, Macmillan Co, 1957

Collins, Randall. Conflict Sociology: Toward an Explanatory Science. New York: Academic Press, 1975.

Coser, Lewis, A. The Functions of Social Conflict. London: Routledge and Kegan Paul, 1956.

Deutsch, Morton, The Resolution of Conflict: Constructive and Destructive Processes. Yale University Press, New Haven, CT, 1973.

Felstiner, William L.F., Richard L. Abel and Austin Sarat. "The Emergence and Transformation of Disputes: Naming, Blaming and Claiming..." Law and Society Review (1980-81): 631-54.

Gamson, William, Power and Discontent, Dorsey Press, Homewood, IL, 1968.

Himes, Joseph, Conflict and Conflict Management, The University of Georgia Press, Athens, GA, 1980.

Mack, Raymond W. and Richard C. Synder, "The Analysis of Social Conflict-Toward an Overview and Synthesis", The Journal of Conflict Resolution, Vol. 1, Number 2.

Ross, M.H. The Culture of Conflict: Interpretations and Interests in Comparative Perspective. New Haven: Yale University Press, 1993.

Ross, M.H. The Management of Conflict: Interpretations and Interests in Comparative Perspective. New Haven: Yale University Press, 1993.

Simmel, G. Conflict. Conflict and the Web of Group Affiliation New York: Free Press, 1955.

Wehr, Paul, Conflict Regulation, Westview Press, Boulder, CO, 1979.

III. SUCCESS IN PUBLIC PARTICIPATION, INTERNAL ORGANIZATIONAL FACTORS, & CONTEXT: IMPORTANT CONSIDERATIONS IN WATER REUSE

Caron Chess

The following remarks draw largely on my research background in public participation because I have limited experience with water reuse issues. The issues below are ones that are increasingly considered in discussions about public participation.

Public Participation Success

One of the problems with discussing public participation cases (or developing guidance for practice) is defining what is meant by successful public participation. For some, successful public participation is defined by the process, that is the characteristics of the means of the participatory effort. Based on this definition, criteria for success may include fairness, representativeness, availability of information, quality of data, facilitation, etc. There are researchers and theorists who argue essentially that the participatory "means" are the "end"—that is, if you have a "good" participatory process, by definition the outcome is successful. Viewed from this perspective, analysis of the participatory processes, not the water reuse policies and initiatives, is critical.

However, for others, the outcome of the process is key to defining success. This perspective decries the emphasis on "warm and fuzzy" effects and sees process as only a means to an end; the results determine whether the participatory means are successful. For water reuse, this might mean successful adoption of a water management plan that included reuse. On the other hand, success could be defined by environmental advocates as defeating water reuse plans that might have negative environmental consequences. Or success could be defined as consensus concerning water management plans.

Therefore, when exploring the issue of participatory success one of the key issues is who gets to do the defining: the sponsoring agency? The participants? Theoretical criteria?

My bias, in general, is the muddled middle ground. I think participatory success should be defined in part by the participants (including the sponsoring agency) and also weighed against norms that might apply across most participatory efforts. In general, I think both process and outcome variables need to be considered. I do not feel that good processes necessarily lead to good outcomes. I even think that it is possible to have good outcomes even with what many would consider poor processes (and, in fact, there are a few cases such as this in the public participation literature).

This question of success makes it difficult for me to evaluate the water reuse case studies. On the face, it would seem that San Diego might be considered less successful than Gwinnett County and San Antonio. But, on what basis do we make that decision?

I think sponsoring government agencies and water utilities probably need to give greater thought to the question of success. What are their goals? The most problematic goals can be the unstated ones, such as implicit assumptions that "good process" will lead to "good outcomes." This often leads to an emphasis on participatory forms (e.g., advisory committees, public meetings, citizen juries), which while important, may be less important than how the participatory forms are used (and for what purpose).

For example, advisory committees have generated ill will and failed to lead to solutions because the sponsoring agency was seeking ratification of proposals rather than "advice." However, in other situations advisory committees have been a means of collaborative problem solving. In short, an advisory committee is merely a tool. The tool can be used expertly or poorly for a range of functions from soliciting ideas to transmitting information. Similarly, although public meetings can devolve into shouting matches, in some situations if properly structured, meetings can also be useful for generating ideas.

It also can be problematic to rely on only one participatory "tool" for large or controversial participatory efforts. Perhaps, avoiding overreliance on advisory committees is one way to deal with the critique (raised by the water reuse cases) that small, multi-stakeholder advisory committees risk isolating representatives from their constituencies' interests and may result in decisions that the constituencies' cannot support.

Instead, a variety of participatory mechanisms may be needed for a variety of audiences. For example, a state agency developing water quality regulations developed an advisory committee that met for a day a month. However, the agency also held informal meetings with a variety of stakeholder groups, developed a thought-provoking "survey" for stakeholders to consider tradeoffs, provided copy for newsletters, mailed updates to mailing lists, etc. In this way, the agency was able to exchange ideas with a range of people—instead of relying on opinions of a few.

Internal Organizational Factors

One of my interests is how internal organizational factors influence public participation. In other words, how does what happens inside an agency influence how it responds to outside stakeholders? An agency (or utility) may set up buffers between key decisionmakers and outside stakeholders, limiting the potential impact of participatory processes. (In these instances, advisory committees can serve to insulate managers from conflict but escalate the level of conflict surrounding the larger organization.)

Context

Citizens in Resolve's water reuse cases did not seem to have initial understanding about sources of water and problems with supply. This conforms to mental models research in New Jersey which suggests that homeowners who relied on public water supplies had less understanding about water supply and management issues than those on well water. Homeowners who dealt with incidents of contaminated water were far more aware of a myriad of issues than homeowners who never had to think about their water. In short, our research underlines the adage "You don't miss your water until the well runs dry."

The question posed by Resolve's cases is: Can you get people to discuss water before their wells run dry? There is ample evidence that it is tough to do. Perhaps, we need to look at other ways to make water issues salient before there is a crisis. For example, close calls in the form of short-term droughts and flooding may be useful prompts.

Problems elsewhere may also serve as stimuli (just as California's energy crisis has prompted discussion in other states). In San Antonio and Georgia, larger water supply issues were explored as part of the planning process that led to recommendations for reuse. Perhaps it makes sense for water reuse to be placed in a larger context of water planning, just as the question of landfills needs to be placed in the larger context of waste management.

IV. PUBLIC PARTICIPATION AND PERCEPTION IN WATER REUSE INITIATIVES

James Conklin

Introduction

The creation of knowledge and consensus, and the translation of consensus into effective strategies and implemented plans, has become a matter of primary concern in the three decades since Peter Drucker proclaimed the imminent arrival of the "knowledge economy." The sustainability and prosperity of both communities and organizations, it is said, will depend significantly on our ability to create processes that permit and encourage disparate groups to collaborate in solving common problems.

With the introduction of new technology and science into the day-to-day lives of communities and organizations throughout North America, much attention has been paid to the factors that contribute to a smooth and safe implementation program. Over the past few decades, a growing awareness has emerged that one of the key factors underlying the successful introduction of a new scientific or technical breakthrough into widespread use is to take explicit steps to promote acceptance and use by a broad spectrum of the public.

Initially, this led to the development of a variety of disciplines collectively referred to as *change management.* Change-readiness assessments are undertaken, communication strategies devised, and the familiar axioms of the 1990s – "communicate, communicate, communicate" comes quickly to mind – are proclaimed in boardrooms and town halls.

The very term *change management*, however, contains within it the seed of its undoing. When dealing with complicated science and technology and diverse communities of stakeholders, it is simply misleading to suggest that social change can be *managed.* Applying proscriptive, top-down management techniques to promote consensus and adoption will usually be seen for what they are – efforts to manipulate a group of people into supporting a new initiative.

More recent approaches tend to focus on providing opportunities to members of a community to participate meaningfully in analyzing problems and developing and implementing solutions. Moreover, research is now being carried out by organizational theorists to consider the processes, systems, and interactions needed to allow groups of people to form relationships, share and create knowledge, and solve problems.

The three case studies and ancillary materials make it clear that most people wish to minimize human contact with recycled water, and to protect both public health and the environment. They also wish to ensure that water conservation measures are an integral part of a forward-moving water strategy, and that the technologies and systems to treat and distribute recycled water must be cost-effective. The materials also suggest, however, that it is unwise to assume that people share a common view on water supply problems. Moreover, it is clear that the logistics surrounding specific water issues can result in what appears to scientists and engineers as a straightforward technical debate become transformed into a broader debate involving basic issues of democratic rights.

The background materials provided for this symposium suggest that there is both significant public concern about the re-use of water, especially in relation to human re-use, there is also a genuine interest in participating meaningfully in helping to guarantee a safe and plentiful supply of water for the future.

Conclusions and Recommendations Based on the Case Studies and Readings

My reading of and reflection on the materials provided suggests compelling links to the work I am seeing carried out on organizational change and collaborative techniques for creating shared understanding and knowledge. I have five points that I would like to share with symposium delegates.

1. An effective and meaningful notion of "participation" must supplant the recent emphasis on manipulative techniques to promote "change management" and "buy-in."

Both theory and practice in organizational change suggest that effective participation involves the creation of trust and collaborative relationships among stakeholders. Participation is not limited to "providing input." Instead, it must clearly include the ability to shape and influence outcomes and to share in the decision-making process. Those responsible for future water re-use initiatives would be well advised to follow processes that permit full and effective participation by all stakeholders.

This means that effort must be made to help stakeholders overcome any knowledge barriers due to specialized technology or complex science. Those in authority must be prepared to share decision-making. A basic model for participative action should include steps to create relationships, communicate and learn, create new understanding and knowledge, and move to action.

2. Participation must begin early, and be continuous throughout the initiative.

Too many change initiatives are undone because the participation of stakeholders is seen as something that can begin at a midway point in the process. Participation should begin at the beginning, and opportunities to participate should continue throughout the initiative. I was especially struck by the Gwinnett County case, where the advisory panel evolved into the advisory board, and a five-year commitment was required. Those responsible for future water re-use initiatives could consider forms of participation that evolve throughout the life of the initiative, from less formal to more formal governance arrangements, and from advisory to decision-making roles.

3. "Facilitation" is a key concept and skill to make participation effective.

Current practice suggests that effective participation involves a "letting-go" of authority by traditional management bodies, so that decision-making power can be shared among a broader stakeholder community. Perhaps one way of grasping this transition is to think in terms of facilitation rather than management. Those with ultimate responsibility for water re-use initiatives will act as the facilitators of a collaborative process, rather than as managers of a top-down process.

In the Gwinnett County case study, a county official said that the process would not have worked if it had been overseen by a professional facilitator. That may be the case; however, my experience suggests that successful facilitation depends on having a facilitator who possesses both thorough knowledge of the situation as well as outstanding facilitation skills. I urge those responsible for future water re-use initiatives to ensure the presence of both (not necessarily in the same person).

4. Translating consensus built through participative activity into follow-on actions.

The case studies tend to focus our attention on the two concepts of participation and perception. They do not, however, touch on challenges that may have arisen in translating plans into action. Research done on organizational strategic planning in the 1980s suggests that the major challenges arise not when creating strategies and plans, but rather when moving to implementation and action.

The question arises: to what extent does a participative planning process ameliorate this challenge. Is the lack of participation one of the barriers to implementing plans? What is the ratio of failure and success for participative planning processes and traditional processes? Before we can fully evaluate the results of participative approaches, some attention must be paid to these questions.

5. More detailed information on specific participative techniques.

The case studies were extremely useful in getting at the issues of participation and perception. However, those responsible for future water re-use initiatives would likely find it helpful to see more detailed and concrete information on participation and collaboration.

What specific governance arrangements were created? How were meetings conducted? What support was available to help stakeholder participants? What collaborative techniques for discussion, idea-creation, planning, and knowledge creation were used? How were documents prepared? Who "owned" the group's intellectual property? How were decisions made?

V. USING "BLUE-RIBBON" OR ADVISORY COMMITTEE

Margot W. Garcia

I used to think that creating a "Blue-Ribbon" Committee to handle a hot topic was a political cop-out. I have since changed my mind. I think that a "Blue-Ribbon" or Advisory Committee, carefully chosen to represent a wide variety of views, including critics, can go a long way in calming down a controversial topic. Their job is to discuss the issue with the best technical advice available, both from the local government as well as from impartial professionals, and write a report with their recommendations. Constructing such a committee has to be done carefully and there are risks for both the members and the local government. The issues that have to handled carefully are: who is asked to serve, who heads the committee, the length of appointment, who sets the agenda, and how the resultant report is issued.

Who is Asked to Serve?

There are any number of ways that the committee can be assembled. I think the best way is for each council member/supervisor to be asked to appoint one member (that builds some political buy-in into the process) and the City/County manger should appoint 3 or 4 people.[In one very hot issue I was involved in, we asked the city manager to appoint all the members because the elected officials were facing a recall/election and we recognized the need for some continuity past the reelection/demise of the elected officials.] The City/County manager has to be sure to balance the political appointments with people who are either the critics (if the elected officials didn't appoint them), leaders of important interest groups (Chamber of Commerce, large local industry, Sierra Club or other large environmental group, neighborhood or civic association council, racial or ethnic groups, and "shakers and movers"), and impartial professionals (either university professors, or members of appropriate professional organizations). The group needs to have a balance of men and women, racial and ethnic minorities, and geographic areas of the city/county. Assembling the group can be quite a task. All the people invited to serve need to be willing to put in the time and effort and recognize that this will be a working group with frequent meetings.

Leadership

The Chair should be appointed by the Mayor/Chair of BOS (Board of Supervisors). This person should be someone committed to making this effort work, have good leadership skills, and be respected throughout the community. The reasons for appointing the chair, as opposed to electing her/him from the group, is to avoid early battles, that everyone may not know each other, and the small group that does know each other and elects one of their own will be seen by the "outsiders" as setting up a clique. If this happens, you have already added another political battle to the group. You don't want that to happen. The focus needs to be on the issue at hand.

Length of Term

Most complex issues will take a year to write a report, even meeting every month. To take only six months will require meeting twice a month. So people should be appointed for the duration of the task. If the group is working together successfully and the elected officials feel it has something to offer over the long term, then the members can be asked if they wish to continue for one or two years and staggered terms should be set up. People should serve no longer than

4 years. Serving longer than that, the group gets stale and becomes a rubber stamp. It needs new blood to ask the question "Why do you do that?" and someone to challenge the assumptions that underlie the decision-making.

Who Sets the Agenda?

The agenda for each meeting should be set by the Chair of the committee and the staff working together. The Chair will know what the members think is important to discuss, the staff will know the resources available for education and will also bring forward topics they think are important. Just having the staff set the agenda makes the committee feel like it is just reacting and has no control over what it wants to do. The Chair needs also to bring to the agenda topics the committee members have asked to be placed there.

Issuing a Report

The purpose of the report is to advise the elected officials on a policy issue. The committee has to remember that it is only issuing advice. The elected officials can adopt it in whole, in part, or not at all.

Who writes the report is always a big issue because the committee members want control over what is written, but often don't have the time to actually do the writing. However, the committee members may have sections they want to write. So, the committee as a whole should set up the outline for the report and see who wants to write what sections. The staff should fill in the rest. The whole report needs to be lightly edited (that means fixing typos and grammar) by staff, but they should be very careful not to change meaning and intent. The entire document needs to be read by the committee and agreed to by vote. If the vote is not unanimous, the negative votes should be allowed to write dissenting opinions – why they could not support the document - and those should be made part of the document when it is released.

There needs to be a press release when the document is finished, copies placed in libraries, on the web, and available to be mailed to anyone who asks. It should be seen as a document for discussion.

Commentary

The reason I changed my mind about the use of the "Blue-Ribbon" committee as described above is because I see it as a useful exercise in getting opinion leaders in the community to hear the same technical information the elected officials have heard, and to work through the issue to a conclusion. It takes away the ability to just shoot from the hip without adequate information and also forces a citizen group to recognize the tradeoffs among different interest groups that the elected official faces. The resulting recommendations can be seen as solutions run up the flag-pole to see who agrees and who fights it. This takes some of the pressure off the elected official. They can sit back and watch the debate, though they will be dragged into it eventually.

The elected officials have to be prepared to get results they don't like. It will be politically expensive for an elected official to vote against the recommendations of the "Blue-Ribbon" committee, so the elected official needs to be careful about what they say on the topic in advance of the "Blue-Ribbon" Committee report. On the other hand, if the elected official stays in touch with her/his representative on the committee, he/she will know which way the committee is leaning and be ready.

For the citizens, the risk is being co-opted by the prestige of serving on the committee, the camaraderie of the group, and the sense that they are making the decision. The committee always has to remember that it is advisory to the elected officials. However, the ones appointed by an elected official have a door open to the person that appointed him/her, and should take advantage of it to keep the elected official informed what is happening and lobby her/him.

The role of outside professionals (that is not local government staff) is important in giving the committee "unbiased" information, using the best science available. The staff has to be prepared to listen and handle different opinions from their own. They can debate with the outside professional in front of the committee and in fact that gives the committee an opportunity to listen to both sides and try to sort things out and come to their own opinions. Often the problem is the underlying assumptions, and that is important for both committee and staff to understand. The staff has to be careful not to undermine or disparage alternative opinions. The Chair's handling of the meeting is very important in these types of situations.

Since trust between citizen and government is often at the base of disagreements over policy, though the disagreement can also be based in fundamental differences in view point, creating a well-balance "Blue Ribbon" Committee that includes critics to discuss the issues can establish trust and go a long way in calming down a hot topic. However, to create the committee without critics, without racial and ethnic minorities, will only make the committee's recommendations subject to even more criticism based on a lack of fairness and opportunity to be heard. For in the long run, the issue does not matter if people feel they have been shut out of the debate. They will fight with the deep seated energy and anger that comes from feeling their basic rights as citizens have been violated.

VI. COMMUNICATION EMBEDS INFORMATION INTO OUR VIEWS OF THE WORLD AND SHAPES HOW WE PERCEIVE WATER REUSE

Troy Hartley

Water is Water . . . Or is It?

Early in my work on water reuse I heard someone say, "Water is water. It doesn't matter what you put in it, whether it's beer or waste." Yet water has social meaning far beyond its simple molecular make-up, H_2O, or its role as a transporter of other things. Socially, it is spiritual in the form of holy water, a national treasure as a waterfall in Niagara Falls or the geysers of Yellowstone, and at times, a fuel for international disputes ripe with historical and cultural conflict. Even in its physical form, water displays such variety that to a casual and intelligent observer, water is many distinct things. It is a glacier or an underground aquifer, a river that starts as a drop in the mountains and flows to a wide delta entering a massive ocean, and even ice on a distant comet or moon. So, in the minds of many, water is so much more than water.

Communication behaves like water in that it can be a transporter of many other things and it shapes much larger physical and social meanings. It can be a simple word or first impression, like a single drop of water. As many communicative acts come together over time and in various venues, it forms a powerful river that can carve and shape the social landscape and relationships between government officials, citizens, elected officials, business leaders, journalists, and others in decision-making processes. In other words, communication is a critical mechanism for shaping our knowledge structure, our sense of right and wrong, and our societal relationships.

Communication Shapes Societal Relationships

From my perspective in political communication and environmental policy, communication must transmit far more than ***scientific or technical information***. The case examples and literature highlight that communication must convey a ***genuine commitment*** that people and their views matter even if I may not understand those views or their immediate relevance. It must get across that decision-making is ***fair*** to all. Communication among participants must take place early, often, and be long-term; this consistency ***builds the trust***, relationships, and credibility among people necessary for communities to make difficult decisions. In fact, without the credibility and trust, the scientific and technical information will carry little weight with many citizens and elected officials. The river of communication begins with a handshake and a first impression, leads to lengthy, sometimes meandering and sometimes raging decision-making processes, and carves the relationships among people and the community's culture.

One scientific view of communication argues that for any type of information (e.g., technical, scientific, values, cultural, etc.) to influence a decision and shape an outcome, it must be embedded in the way decision-makers think and frame the questions and solutions they are considering. To embed information, it must be discussed and receive focused attention – that is, a cyclical process of effective articulation of ideas, active listening, skillful questioning and probing of views, serious reflection on ideas and practice, and the generation and communication of new and innovative ideas. There seems to be many form of deliberation that accomplish this embedding of information; however without this level of deliberation ***new*** ideas and information that are counter to people's existing knowledge structure and experiences, like water reuse, will not be understood or supported by decision-makers, particularly citizens and elected officials. Messages that are consistent with their existing knowledge structure and experiences, like "toilet to tap," will be far more easily accepted because the idea is already embedded in the way people think. In other words, if you want to convince someone to accept an innovative idea that is not consistent with their existing view of the world (or view of water), then it will be critical to engage in a high quality public dialogue.

Communicative Lessons in the Cases & Literature

Below I apply these perspectives on the cases and the literature. Features of the public dialogue and the nature of communication in the various cases, raise several possibilities for communicative lessons learned in regards to water reuse:

- The public involvement strategies that seem to be most successful convey a **genuine commitment** to hear everyone out, capture their opinions even if they are dissenting views, and respectfully discuss people's concern. One way this commitment can be expressed is through government and organizational leaders endorsing and championing a decision-making process. It can also be conveyed by individuals who show respect for others' views, promote an honest approach to communicating information and views, and are not afraid to admit mistakes. Communicating a genuine commitment can be slightly complicated by statutory or administrative authority granting one government agency the final decision-making authority. In this case, officials need to be adept at sharing decision-making power, while retaining authority. Finally, people seem to be able to detect when communication is not genuine and it makes them suspicious and quick to reject or oppose information.

- To date, the best **motivation and incentives** for initiating water reuse seems to be a widely acknowledged crisis. There can be many sources of that crisis, all of which are communicated in many ways on a daily basis – e.g., drought, growth and development, imported water or underground aquifer less likely to continue to supply same volumes of water. It is less clear that an initiative can be successful without a crisis, although it would certainly make for better public policy. One of the challenges to undertaking a potable water reuse initiative would be to address the interests of enough parties, or enough powerful parties (e.g., the consumer market for reused water in the San Antonio case) to ensure their continued participation. This suggests that a broader water resource management discussions that consider both drinking water and wastewater may more likely attract participants, permit long-term discussion, and thus build the trust, credibility, and potentially acceptance of some form of potable water reuse.

- **Nonpotable and indirect potable water reuses** *may* be perceived in fundamentally different ways by the public, although this is a hypotheses based upon very few cases and discussions over the past year with water reuse professionals. The two forms of reuse could be like apples and oranges to the public, with non-potable uses being widely accepted and indirect potable initiatives often being rejected. This suggests that the nonpotable use is consistent with the way the public currently frames in their mind how treated wastewater can and should be used. It also suggests that the more frequently people encounter and support non-potable reuses and thus, enhance their knowledge structure and experience with it, the less likely they may be to endorse potable uses because it is counter to their pre-existing knowledge structure. Further, if this is true, then an indirect potable reuse project may actually be more successful in an area with no history of non-potable reuse, like Gwinnett County, than in an area with a long history, like San Antonio.

A Challenge Identified Through the Literature Review

Several challenges for conducting public participation and for building trustworthy, credible, and scientifically sound public perceptions emerged from my reading of literature. One in particular, I will discuss here, the **nature of representation** in various participation mechanisms. Representatives in public participation processes (e.g., Citizen Advisory Boards) are often leaders or delegates from an interested stakeholder group. While groups of negotiators may interact and discuss issues enough to develop the trust and mutual understanding necessary to reach difficult agreements, the constituencies of the representatives often do not. They may not share the trust and understanding that their representatives do, and consequently may not support the agreements. In fact there are a few people who strongly oppose the use of collaborative, consensus-building processes because of this phenomenon.

However, rather than discarding the small advisory groups as a viable public participation mechanism, I believe that we need an expanded view of the representative. In addition to their role as a voice for a particular interest, I believe that participants on small advisory groups have a duty and obligation for the privilege of being a representative in a democratic decision-making exercise. That duty demands that they consistently and adequately conduct outreach to their constituency, to remain the credible voice for the constituency and ensure that they have the most up-to-date views of their constituency. It is not enough to be an adequate negotiator, a representative must be a skilled public education and outreach coordinator.

In addition, it is in the best interest of government participants and others to ensure that the representatives are serving these roles to keep the constituencies abreast of decision-making progress and carry the enhanced trust, credibility and legitimacy back to the broader constituencies. Active support of representatives could involve using technical assistance grant funds for outreach and education expertise. No issue should emerge from a constituency at the 11th hour to surprise an advisory group. Last, government sponsors should make certain that the advisory group membership roughly parallels the demographics of the community, to ensure that under-represented interests (e.g., environmental justice) do not emerge late in the process.

Conclusion

From my perspective, water reuse, particularly indirect potable uses, would need to overcome a strong negative perception that is currently embedded in the way people think about the issue and the solutions. This perception emerges from consistent communicative messages that bombard us daily from many sources. To modify existing perceptions, it is necessary to engage in prolonged discussion that builds trust among people with diverse perspectives. The discussion must be seen as fair and address all points of view and information. It must be genuine and honest, and people must be willing to engage in it for the long haul. Without embedding an alternative perception, short catchy phrases that play to the existing knowledge structure like "toilet to tap" will always win out over scientific and technical arguments.

VII. COMMENTS ON, "PUBLIC PERCEPTION AND PARTICIPATION IN WATER REUSE: LITERATURE SUMMARY"

Mary McDaniel

Why Can't We All Just Get Along?

Like most practitioners in the trenches, I am often frustrated by the nature of discourse over controversial siting, clean-up, and public work projects. I have observed behavior on both sides of the table that makes me cringe even now. Discourse is often dominated by a few true believers on both sides, often with extreme views, and is usually competitive and antagonistic. Rather than launch into a diatribe about the lack of polite discourse, I would like to comment on the difficulty of getting people involved in the first place, and the even greater challenge of keeping them involved through protracted and often tedious bureaucratic processes.

Peter Sandman and many others have described the natural state of the majority of the populace as apathetic. A common attitude is, "If it doesn't affect me personally right now, I'm not interested." It is hard to argue with this point of view, although I think the term "apathetic" is somewhat pejorative. The complexities of modern life are enough to occupy one's time without an endless parade of public meetings, particularly when they deteriorate into the equivalent of a bar room brawl.

For recycled water projects, there is the additional element of making people face an issue that most would rather not address at all. There is no such thing as "new" water. However, when asked to deal with the idea that the water they are drinking today could have been in Ceasar's kidney's, people are extremely uncomfortable. Most people do not know where their drinking water comes from and most do not want to know, as long as it is not limited and is perceived to be safe. When faced with the specter of viruses, bacteria, chemical contaminants, byproducts of industrial operations, and pharmaceutical residues, many would choose blissful ignorance.

I think that, in most cases, talk of "stakeholder involvement" is no more than lip service. There is still a large measure of what has been dubbed as, "decide, announce, defend", the parochial "DAD" model of decision-making, being used in the real world. Lots of pointy headed white male engineers completely design a project then decide to hold a public meeting for "shareholder involvement" and do not understand why people are angry because they have no control over the decision-making process.

I do not have a magic bullet for always getting the process right. However, I do know some sure fire ways to make it fail.

1. Make all the decisions first.
2. Be inflexible.
3. Be condescending.
4. Skip the refreshments ("People will think we are trying to buy them with a cookie.")
5. Have all your presenters wear suits and ties and sit on a stage behind a draped white table.
6. Make everyone hold their questions until the end of a one and a half hour presentation.
7. Be defensive. Be technical.
8. Offer the euphemism "secondary effluent" in place of treated sewer water. Hope they won't make the connection.

9. Don't follow-up.
10. Forget to turn your cell phone off.

I believe that most people want to trust the agency that is managing their drinking water. However, trust is built on a personal level over time and on a consistent track record. When a public participation process is a hasty add-on, there is little time to develop the trust and credibility that allows for a true and productive dialogue.

Most people acquire their health, environment, and safety information from the media. However, if agencies are nervous about inviting public participation, they are doubly nervous about involving the press. Again, the process of building a relationship with the press takes time. Reporters often have a very limited time to understand very complex projects when they have deadlines and only a few minutes to become "instant experts". The agency should then not be surprised if the issue is erroneously reported.

In my experience, there are a number of key ways to motivate the public to become involved in dialogue with an agency. These include:

- Involve the public early, before all decisions have been made.
- Understand the psychological significance of the issue to the public. For example, agencies often underestimate the symbolic significance water has for their constituents. Water represents innocence and telling them where it comes from is a fall from grace. Respect the public. Recognize the importance of "fairness" both in substantive discussions such as siting, as well as in process.
- Empathize. Empathize. Empathize.
- Share control. Allow for questions. Do not stick to your agenda. Listen and take notes.
- Take an interest. Talk with people. Make yourself a little vulnerable.
- Be available. Give people contact information and encourage them to use it.
- Follow up all requests for additional information.
- Remember this is a process and keep people updated.
- Encourage tours to take the mystery and the "yuck factor" out of the treatment process.
- Involve the press. Present the human interest side to the issue. Provide graphics and backgrounders.

By way of summary, I would like to quote an excellent passage from a document published by the Water Environment Federation, entitled Using Reclaimed Water to Augment Potable Water Resources:

"The ability to gauge the full scope of underlying issues is also critical to a project's final acceptance by the public. Although technological considerations may open the door to dialog, leading a community through the door requires mobilizing interest and thoughtfully managing the process. Bare facts and technical assurances may not guarantee the success of a potable reuse project. Although there are tricks of the trade such as avoidance of jargon and preparedness, a process may also require basic character strengths of patience, empathy, respect, wisdom, sincerity, and humor. The ability to apply these characteristics comes from stamina and experience." (WEF, 1998, p. 231)

VIII. PROMOTING POTABLE REUSE IN A DIFFICULT PUBLIC/POLITICAL ENVIRONMENT

Michael J. McGuire and Edward G. Means

The literature review and case studies prepared for this project demonstrate how difficult it is for potable reuse projects to be accepted by the public. However, no matter how difficult it is, public officials must address the issues squarely and find ways to deliver successful projects. Using conservative fertility and immigration rates, the population of the U.S. is expected to double by the year 2100. With a finite water supply and expensive desalting alternatives, we have no choice but to plan innovative and ambitious indirect potable reuse projects.

If future indirect potable reuse projects are to be successful, we cannot tolerate any more public acceptance failures. The public and the political establishment will dismiss out-of-hand any future potable reuse proposals unless a success story can be implemented. Obviously, the key question is how do we get the first success?

As a result of our participation in some of the most aggressive potable reuse projects in California, we believe that the following points, if adopted, will help make acceptance of that first project possible.

One fact that is rarely discussed in this arena is the complete acceptance in California by the public and political establishment of existing potable reuse projects. If there was a total abhorrence of potable reuse, there would be a clarion call to dismantle these projects. In addition, unplanned potable reuse is occurring on all major river and lake systems in the U.S. There is tacit public approval of this situation. We have to understand why people accept these "grand fathered" projects and why they reject the new projects that very often have far more pathogen barriers and overall treatment reliability. We also have to stop being afraid to call to the attention of the public in a forceful manner, that the water they are currently drinking is reused already.

It is clear to us that any future potable reuse project in any city or region must be led by a governmental unit other than the wastewater utility. A fundamental flaw in the San Diego Repurification Project was its transfer from the San Diego County Water Authority (a regional water utility) to the wastewater department under the City of San Diego. In some cases a new entity should be created to spearhead the effort to employ indirect potable reuse as a water supply augmentation effort. If we are ever to break the flip phrase "toilet-to-tap," we have to get the "toilet" agency out of the leadership role.

In this age of instant communication, it is also clear that the message for any future reuse project must be carried by the Chief Executive Officer of the responsible entity. Using technical staff to lead these efforts has lead to problems and miscommunications because the technical staff rarely see the big picture including the political side. The acceptance of Colorado River water in Tucson, Arizona this year after a debacle over the period 1992-1994, was accomplished in great part by the immersion of the Director of Tucson Water, David Modeer, in the public process. He was clearly identified as the person responsible for making the project happen without further disaster.

The message that is delivered by the lead utility must be focused and deal with the issues crucial to the acceptance of potable reuse. Once again drawing from our experiences in Tucson, it was not until Tucson Water made it clear through a huge outreach effort that the falling groundwater levels was real and potentially catastrophic that the public began to accept the concept of an alternative, surface supply.

The many models detailed in the reuse literature review appear to be a confusing array of choices. It is our view that utilities promoting a reuse project not choose between the options but employ all of the options, or as many as can be afforded. For example, utilities should consider not only engaging a stakeholder group but they should also entertain the use of a randomly selected citizens panel. In addition, a panel of experts can be useful to review the technical soundness of the project. By itself, the technical panel will not convince everyone, but it is an absolutely critical component of any overall approach.

Whatever approaches are used, the responsible utility must be totally committed to the process and be ready to be involved in the processes over many years. Bringing on a new indirect potable reuse project will be very expensive and will take tremendous focus and dedication by the utility. The media, public and political participants can tell fairly quickly if a utility is engaging in a public process to just be able to do what they wanted to do in the first place. An attitude of public process as a "necessary evil" will kill any project and is especially deadly for potable reuse projects.

Many of the reuse public processes have employed a Citizens Advisory Committee. We propose that a Political Advisory Committee is at least as important. If the water resources picture is truly as bleak as a utility describes and potable reuse is a critical part of an assured future supply, it is in the best interest of local politicians to be get involved. Very often it is difficult to get their attention but every effort must be made to have them part of the solution. For the San Diego Repurification Project, local politicians became the problem and foiled any solution.

With whatever public processes are chosen, the utility trying to implement the project will have to share power in a clear and meaningful way. Once again, "empowerment" as window dressing will be detected by all participants early in the game.

Finally, the lead agency or utility must commit to providing a mountain of data to the public and to the individuals involved in the various processes in a manner that is both accessible and understandable. Tucson Water has pioneered a detailed web-based system that provides water quality data using a GIS-based interactive database. To view the future of how water quality data will need to be provided to customers, go to: http://www.ci.tucson.az.us/water/water_quality/water_quality.htm.

IX. ENSURING PUBLIC CONFIDENCE IN RECYCLED WATER PROGRAMS

Jeff Stone

Public perception and acceptance of recycled water programs rests largely on the success of the reuse practices (e.g. – providing technological/scientific assurances of its safety to the general public, ensuring no 'mishaps' with inadequately treated recycled water delivered to use-sites, no cross-connection incidents, no aesthetic problems associated with the recycled water, etc.), which instill public confidence in its use. Public confidence is generally only gained when you have all 'positives' and no 'negatives' associated with the reuse. Anything in the negative category, regardless of how minor it may be, will generally be exploited by the media and general public which results in nearly irreparable harm to a water reuse program.

These issues were given considerable attention by a working group assembled at a Nominal Group Technique (NGT) Workshop sponsored by the National Water Research Institute in 1999. Elements that came to light through this effort are summarized below.

It might be helpful to establish safety equivalency relationships among public health criteria and analogous public activities to expand public education/awareness of the rationale of the reuse criteria and best management practices, and their application and use to ensure the safety of non-potable reuse. This can be conducted in much the same way as analogies are drawn to alleviate concerns with airline travel (how do safety factors associated with recycled water use compare with other common risks associated with our daily lives?). When put into proper perspective with the multitude of factors we face daily that have a potential impact on our own well being, fears are often allayed. Many viable projects have failed, at least in part, not on insufficient science or technology, but on public misperception of their safety.

Effective communication programs are needed that use descriptive terminology to explain treatment technologies, monitoring system requirements and results, reliability standards, etc., in clear, layperson terms which will improve understanding, support and confidence in reuse programs by the general public. The public oftentimes has no idea of the history behind the evolution of water reuse, the demonstrated effectiveness and reliability of the technologies utilized today, the risk assessment studies that have been conducted, the permitting requirements imposed on the producer, distributor and end users, etc. Also proving to be effective is to factor the "green issue" into the equation, as this often casts a positive light on reuse; conserving limited fresh water supplies for higher order uses (i.e. potable supplies), maximizing a finite resource, taking something considered to be a waste (sewage) and turning it into a useful commodity, preventing or minimizing the discharge of 'wastes' into the environment, etc.

Effective communication of the issues to the general public may not always be carried out in the proper manner by those in the industry, as they are not necessarily trained in public relations. A bridge between experts in industry and public relations professionals is needed to develop an effective public education, awareness and confidence building program. It would be useful to bring the scientific, regulatory, industry and public relations groups together to develop an effective approach to developing guidance, which clearly defines the multitude of safety issues (e.g. risk, treatment level/reliability, monitoring, use-site controls, etc.) considered for a project to be permitted.

X. PUBLIC PERCEPTION AND PARTICIPATION IN WATER REUSE DISCUSSION PAPER

CHRIS J. WIANT

Reclaimed water, whether for potable or non-potable use, is a topic that often evokes a strong and emotional response among those who may be asked to utilize the water. As with many perceived involuntary risks the degree to which individuals will respond seems to be directly related to the probability for human consumption of the reclaimed water. It is a virtual certainty, in the U.S., that the quantity of water provided to consumers through direct reuse will increase. Furthermore, the challenge of effective water treatment continues to become more complex as demand for available surface water sources increases and as they receive more discharges from wastewater treatment plants, storm events and other sources. A better public understanding and acceptance of the alternatives to meeting the public demand for water that also ensure the availability of a water supply that protects public health is imperative if the full value of reclaimed water is to be recognized.

The case studies presented from San Antonio, San Diego and Gwinnett County illustrate a few of the subtle but critical issues associated with the public's perception about water reuse and the nature of their participation in the decision making process. The case studies are also instructive for those who may engage in efforts to support water reuse projects in other cities. Following are brief descriptions of some of the issues that may generate further discussion.

The case studies illustrate that assumptions about the public's understanding of water reuse were made early in the process by those charged with managing public involvement. For example, in San Antonio a survey showed little buy-in to a problem of water quality and quantity in the community water supply. The public focused on minimizing lifestyle disruptions. Will they be willing to make sacrifices (reduce consumption) or endorse alternatives (water reuse) if they have not accepted that a problem exists? It is prudent to test the assumptions that are made, especially as they relate to the public's perception of the problem.

In each case one could speculate that the public might view a contemporary drinking water supply source as free from the contaminants that may be present in reclaimed water. However, discharges from wastewater treatment plants are allowed into surface waters that may become the source of a downstream community water supply. Without some fundamental understanding of the quality of their current water source the public may perceive they are going from a pristine source to a contaminated source of water. The public's understanding that the critical factor in the quality of potable water is an effective and quality assured treatment process is important. Poor quality source water can be made safe through effective treatment.

Two case studies, San Diego and Gwinnett County, also implied that the preferred alternative to resolve the water supply problem was identified before the public participation process began. The panel was asked to review the preferred alternative without having participated in discussion that evaluated the alternatives. San Diego officials relied on the endorsement of the department of health to validate the potable reuse alternative as safe and effective. This issue may have created less controversy where potable reuse was not yet being contemplated. However, when human consumption is anticipated community acceptance of both the problem and the solution is particularly critical. Contemporary literature suggests that there is need for the public and the policymakers to be in the same place in the several steps of coming to public judgment on a given topic. For instance, if policymakers are ready to make a decision and the public is not convinced there is a problem they may protest acceptance of the preferred policymaker alternative. There are also the issues of trust and credibility. Will the public accept the health department's endorsement as credible?

The case studies also illustrate how other agendas can influence the debate on water reuse. For example, the water quality issues in the reservoir in Gwinnett County appear to have generated other associated aesthetic and ecological concerns for the public. Those issues will ultimately have to be addressed in order to satisfy the public participants and enable their consent to move the reuse issue forward. A controversial issue (water reuse) can be used as leverage to achieve other goals (ecological protection) that a community is committed to.

Another topic to consider for each case study is the effectiveness of the outreach efforts to the general public. There was little indication that the opportunity to educate the public about water quality and quantity issues was exploited. There appeared to be marketing strategies to sell the project but not further effort to gain a better understanding and acceptance of water reclamation. It is only a matter of time before San Antonio and Gwinnett County will be looking to accomplish potable reuse. It is anticipated that the public's response in those cases will be similar to that of San Diego, despite the "buy-in" that was achieved in the other communities for non-potable reuse.

In summary, it is easy to focus on the technology issues associated with water reuse. Agencies often conclude that they merely need to get the public to understand the relevant science. However, these case studies effectively show of the importance of understanding how the affected public perceives an issue and their need to participate in the decision making process, just as the responsible agencies do.

XI. AUTHORS' AFFILIATIONS

Maureen E. Austin, Ph.D. is an environmental psychologist at the University of Michigan, School of Natural Resources & Environment, Ann Arbor, Michigan.

Juliana Birkhoff, Ph.D. is an independent scholar and consultant in Vienna, Virginia and former Director of the Center for Research and Education at RESOLVE, Inc., in Washington, DC.

Caron Chess, Ph.D. is Director of the Center for Environmental Communication and an Associate Professor in Rutgers University, Cook College's Department of Human Ecology, New Brunswick, New Jersey.

James Conklin is founder and President of Information Development & Training, Inc. (IDT) in Winnipeg, Manitoba, Canada.

Margot W. Garcia, Ph.D. is an Associate Professor in the Urban Studies and Planning Department at Virginia Commonwealth University, Richmond, Virginia.

Troy Hartley, Ph.D. is a senior researcher, trainer, and mediator at RESOLVE, Inc., a non-profit environmental consensus-building organization in Washington, DC.

Mary McDaniel, DO is a board-certified physician, licensed attorney, and co-founder of McDaniel/Lambert, a risk and crisis management consulting firm in Los Angeles, California.

Michael McGuire, Ph.D. is founder and President of McGuire Environmental Consultants, Inc., and the former Assistant General Manager of the Metropolitan Water District of California.

Edward Means is Vice President at McGuire Environmental Consultants, Inc., and former Deputy General Manager and Chief Operating Officer with the Metropolitan Water District of Southern California, Los Angeles, California.

Jeffery Stone is Chief of the Recycled Water Unit in the Division of Drinking Water, California Department of Public Health, Sacramento, California.

Chris J. Wiant, Ph.D. is President of Caring for Colorado Foundation, and the former Executive Director of the Tri-County Health Department, Denver, Colorado.

REFERENCES

American Chemical Council. 2001. cmahq.com

Austin, M.E. 2001. Public Perception and Participation in Water Reuse Initiatives: Perspectives and Recommendations. In Hartley, ed., 2001, *Public Perception and Participation in Water Reuse: White Papers*, July. Washington DC: RESOLVE Inc.

Birkhoff, J. 2001. Community Conflict Over Water Reuse: Perspectives from Conflict Analysis and Resolution. In Hartley, ed., 2001, *Public Perception and Participation in Water Reuse: White Papers*, July. Washington DC: RESOLVE Inc.

Bingham, G. 2000. Principles of Collaboration and Consensus Building on Public Issues, presentation materials. Washington DC: RESOLVE Inc.

Bingham, G. 1998. "Stages in Consensus Building." presentation materials. Washington DC: RESOLVE Inc.

Broad, P.A. 1996. *Community Views on Water Reuse*, June. Sydney, Australia: Sydney Water Corp., Ltd.

Bruvold, W.H. 1981. Public Participation in Environmental Decisions: Water Reuse. *Public Affairs Report: Bulletin of the Institute of Governmental Studies*, February, University of California – Berkeley CA, 22(1):1-6.

Bruvold, W.H. 1988. Public Opinion on Water Reuse Options, *J. Water Pollut. Control Fed.* 60(1):45-50.

Bruvold, W.H. 1991. *Public Evaluation of Municipal Water Reuse Options: Technical Completion Report*, August. Berkeley, CA: University of California.

Blue Ribbon Panel, 1998. *Findings and Recommendations Related to the Scientific and Technical Review of the City of San Diego Waste Repurification Project*, September 16. San Diego, CA.

Chess, C. 2001. Success in Public Participation, Internal Organizational Factors and Context: Important Considerations in Water Reuse. In Hartley, ed., 2001, *Public Perception and Participation in Water Reuse: White Papers*, July. Washington, DC: RESOLVE, Inc.

Conklin, J. 2001. Public Participation and Perception in Water Reuse Initiatives. In Hartley, ed., 2001, *Public Perception and Participation in Water Reuse: White Papers*, July. Washington, DC: RESOLVE, Inc.

Garcia, M.W. 2001. Using "Blue-Ribbon" or Advisory Committees. In Hartley, ed., 2001, *Public Perception and Participation in Water Reuse: White Papers*, July. Washington, DC: RESOLVE, Inc.

Hartley, T.W. 2001. Communication Embeds Information Into Our Views of the World and Shapes How We Perceive Water Reuse. In Hartley, ed., 2001, *Public Perception and Participation in Water Reuse: White Papers*, July. Washington, DC: RESOLVE, Inc.

Huckin, T., and L.A. Olsen. 1991. *Technical Writing and Professional Communication*, 3-18. New York: McGraw-Hill.

International Association for Public Participation (IAP2). 2000. *IAP2 Public Participation Spectrum*, www.iap2.org, Denver CO: IAP2.

International Association for Public Participation (IAP2). 2000. *IAP2 Public Participation Toolbox*, www.iap2.org, Denver CO: IAP2.

Jeffery, P. 2001. *Understanding Public Receptivity Issues Regarding "In-House" Water Recycling: Results from a UK Survey*. Unpublished manuscript, Cranfield University UK.

Johnson, R.J., and M.J. Scicchitano. 2000. "Uncertainty, Risk, Trust, and Information: Public Perceptions of Environmental Issues and Willingness to Take Action," *Policy Studies Journal*. 28(3):633-647.

Kaplan, S, and R. Kaplan. 1982. *Cognition and Environment: Functioning in an Uncertain World*. New York: Praeger.

Kaplan, S, R. Kaplan, and R.L. Ryan. 1998. *With People in Mind: Design and Management of Everyday Nature*. Washington, DC: Island Press.

King, C.S., K.M. Feltey, and Susel B. O'Neill. 1998. "The Question of Participation: Toward Authentic Public Participation in Public Administration," *Public Administration Review* 58(4):317-326.

Lawrence, G. 2000, *Survey of Orange County Voters Within the Orange County Water District*, December. Santa Ana CA: Lawrence Research – Public Affairs & Marketing.

Lohman, L.C., and J.G. Milliken. 1984. Public Attitudes Toward Potable Wastewater Reuse: A Longitudinal Case Study. In *Proceedings of the Water Reuse Symposium III: Future of Water Reuse, Volume 1*, San Diego CA, August 26-31, pp. 109-121.

McDaniel, M. 2001. Comments on *Public Perception and Participation in Water Reuse: Literature Review*. In Hartley, ed., 2001, *Public Perception and Participation in Water Reuse: White Papers*, July. Washington, DC: RESOLVE, Inc.

McGuire, M., and E. Means. 2001. Promoting Potable Reuse in a Difficult Public/Political Environment. In Hartley, ed., 2001, *Public Perception and Participation in Water Reuse: White Papers*, July. Washington, DC: RESOLVE, Inc.

Moore, C.W. 1987. *The Mediation Process: Practical Strategies for Resolving Conflict*. San Francisco: Jossey-Bass Publishers.

National Research Council (NRC). 1998. *Issues in Potable Reuse: The Viability of Augmenting Drinking Water Supplies with Reclaimed Water: Committee to Evaluate the Viability of Augmenting Potable Water Supplies with Reclaimed Water*, March 11. Washington DC: National Research Council, www.nationalacademies.org.

The Pew Research Center for the People and the Press. 2001. http://www.people-press.org/questionnaires/trustque.htm.

Putnam, R.D. 1995. "Bowling Alone: America's Declining Social Capital," *Journal of Democracy* 6(1):65-78.

Rabe, B.G. 1994. *Beyond NIMBY: Hazardous Waste Siting in Canada and the United States*. 1994, Washington, DC: The Brookings Institution.

Renn, O., T. Webler, and P. Wiedermann. 1995. *Fairness and Competence in Citizen Participation: Evaluating Models for Environmental Discourse*. Boston: Kluwer Academic Publishers.

Stern, P.C. 2001. National Resource Council's Study of Public Participation in Environmental Assessment and Decision Making notes and workshop meetings, Washington DC: National Academy of Sciences/National Research Council, Commission on Behavioral and Social Sciences and Education, Committee on the Human Dimensions of Global Change.

Symposium. 2001. Framework for Public Perception and Participation in Water Reuse Initiative Stakeholder Symposium, August 2-3. Washington, DC: RESOLVE, Inc.

U.S. Environmental Protection Agency. 1992. *Manual: Guidelines for Water Reuse*, Office of Water and Office of Research and Development, EPA/625/R-92/004, September. Washington DC: U.S. Environmental Protection Agency.

Water Environment Federation. 1998. *Using Reclaimed Water to Augment Potable Water Resources*. Arlington, VA: Water Environment Federation and the American Water Works Association.

WEFTEC. 2001. Public Perception and Participation Workshop, Water Environment Federation 74th Annual Conference and Exposition, October 13-17, Atlanta, GA.

Wildermuth, R. 2001. Groundwater Replenishment System: Public Outreach and Education Program. Presentation at the NWRI National Urban Watersheds Conference, October 17-19, Costa Mesa, CA

Wilson, J.Q. 1989. *Bureaucracy: What Government Agencies Do and Why They Do It.* New York: Basic Books.

Yearly, Steven. 2000. Making Systematic Sense of Public Discontents with expert Knowledge: Two Analytical Approaches and a Case Study, *Public Understanding of Science* 9(2):105-122.

2003 WERF SUBSCRIBERS

Wastewater Utility

Alabama
Montgomery Water Works & Sanitary Sewer Board

Alaska
Anchorage Water & Wastewater Utility

Arizona
Gila Resources
Glendale, City of, Utilities Department
Mesa, City of
Phoenix Water Services Department
Pima County Wastewater Management

Arkansas
Little Rock Wastewater Utility

California
Calaveras County Water District
Central Contra Costa Sanitary District
Contra Costa Water District
Crestline Sanitation District
Delta Diablo Sanitation District
Dublin San Ramon Services District
East Bay Municipal Utility District
El Dorado Irrigation District
Escondido, City of
Fairfield-Suisun Sewer District
Irvine Ranch Water District
Las Virgenes Municipal Water District
Lodi, City of
Los Angeles, City of
Los Angeles County, Sanitation Districts of
Napa Sanitation District
Orange County Sanitation District
Palo Alto, City of
Riverside, City of
Sacramento Regional County Sanitation District
San Diego Metropolitan Wastewater Department, City of
San Francisco, City & County of
San Jose, City of
Santa Barbara, City of
Santa Rosa, City of
South Bayside System Authority
South Orange County Wastewater Authority
Union Sanitary District
West Valley Sanitation District

Colorado
Boulder, City of
Colorado Springs, City of
Littleton/Englewood Water Pollution Control Plant
Metro Wastewater Reclamation District, Denver

Connecticut
New Haven, City of, WPCA

District of Columbia
District of Columbia Water & Sewer Authority

Florida
Broward, County of
Fort Lauderdale, City of
Gainesville Regional Utilities
JEA
Kissimmee, City of, Department of Water Resources
Miami-Dade Water & Sewer Authority
Orange County Utilities Department
Orlando, City of
Reedy Creek Improvement District
Seminole County Environmental Services
St. Petersburg, City of
Stuart Public Utilities
Tallahassee, City of
Tampa, City of
West Palm Beach, City of

Georgia
Atlanta Wastewater Services
Augusta, City of
Clayton County Water Authority
Cobb County Water System
Columbus Water Works
Fulton County
Gwinnett County Department of Public Utilities
Macon Water Authority
Savannah, City of

Hawaii
Honolulu, City and County of

Idaho
Pocatello, City of

Illinois
American Bottoms Wastewater Treatment Plant
Dupage, County of
Greater Peoria Sanitary District
Kankakee River Metropolitan Agency
Metropolitan Water Reclamation District of Greater Chicago
Wheaton Sanitary District

Indiana
Fort Wayne, City of
Indianapolis, City of

Iowa
Ames, City of
Des Moines Metro Wastewater Reclamation Authority

Kansas
Johnson County Unified Wastewater Districts
Unified Government of Wyandotte County/Kansas City, City of

Kentucky
Louisville & Jefferson County Metropolitan Sewer District

Louisiana
Sewerage & Water Board of New Orleans

Maine
Bangor, City of
Portland Water District

Maryland
Anne Arundel County Bureau of Utility Operations
Howard County Department of Public Works
Washington Suburban Sanitary Commission

Massachusetts
Boston Water & Sewer Commission
Upper Blackstone Water Pollution Abatement District

Michigan
Ann Arbor, City of
Detroit, City of
Grand Rapids, City of
Holland Board of Public Works
Lansing, City of
Owosso Mid-Shiawassee County WWTP
Saginaw, City of
Wayne County Department of Environment
Wyoming, City of

Missouri
Independence, City of
Kansas City Missouri Water Services Department
Little Blue Valley Sewer District
Metropolitan St. Louis Sewer District

Nebraska
Lincoln Wastewater System

Nevada
Henderson, City of

New Jersey
Passaic Valley Sewerage Commissioners

New York
New York City Department of Environmental Protection

North Carolina
Cary, Town of
Charlotte/Mecklenburg Utilities
Durham, City of
Metropolitan Sewerage District of Buncombe County
Orange Water & Sewer Authority

Ohio
Akron, City of
Butler County Department of Environmental Services
Columbus, City of
Metropolitan Sewer District of Greater Cincinnati
Northeast Ohio Regional Sewer District

Oklahoma
Tulsa, City of

Oregon
Clean Water Services
Eugene/Springfield Water Pollution Control
Water Environment Services

Pennsylvania
Philadelphia, City of
University Area Joint Authority, State College

South Carolina
Charleston Commissioners of Public Works
Mount Pleasant Waterworks & Sewer Commission
Spartanburg Sanitary Sewer District

Tennessee
Cleveland, City of
Knoxville Utilities Board
Murfreesboro Water & Sewer Department
Nashville Metro Water Services

Texas
Amarillo, City of
Austin, City of
Dallas Water Utilities
Denton, City of
El Paso Water Utilities
Fort Worth, City of
Gulf Coast Waste Disposal Authority
Houston, City of
San Antonio Water System
Trinity River Authority

Utah
Salt Lake City Corporation

Virginia
Alexandria Sanitation Authority
Arlington, County of
Fairfax County Virginia
Hampton Roads Sanitation District
Henrico, County of
Hopewell Regional Wastewater Treatment Facility
Loudoun County Sanitation Authority
Lynchburg Regional WWTP
Prince William County Service Authority
Richmond, City of

Washington
Edmonds, City of
Everett, City of
King County Department of Natural Resources
Seattle Public Utilities

Wisconsin
Green Bay Metro Sewerage District
Madison Metropolitan Sewerage District
Milwaukee Metropolitan Sewerage District
Racine, City of
Sheboygan Regional Wastewater Treatment
Wausau Water Works

Australia
Sydney Water Corporation
Water Corporation of Western Australia

Canada
Toronto, City of, Ontario

Mexico
Servicios de Agua y Drenaje de Monterrey, I.P.D.

United Kingdom
Yorkshire Water Services Limited

Stormwater Utility

California
Monterey, City of†
Santa Rosa, City of†

Colorado
Boulder, City of†

Georgia
Griffin, City of†

Kansas
Overland Park, City of†

Kentucky
Louisville & Jefferson County Metropolitan Sewer District†

Maine
Portland Water District†

North Carolina
Cary, Town of†
Charlotte, City of/Mecklenburg County†

Pennsylvania
Philadelphia, City of†

State

Fresno Metropolitan Flood Control District, Calif.†
Urban Drainage & Flood Control District, Colo.†

Corporate

ADS Environmental Services
The ADVENT Group Inc.
Alan Plummer & Associates
Alden Research Laboratory
Alpine Technology Inc.
Aquateam–Norwegian Water Technology Centre A/S
BaySaver Inc.†
BioVir Inc.
Black & Veatch
Boyle Engineering Corporation
BPR CSO
Brown & Caldwell
Burns & McDonnell
CABE Associates Inc.
The Cadmus Group
Camp Dresser & McKee Inc.
Carollo Engineers Inc.
Carpenter Environmental Associates Inc.
CDS Technologies Inc.
Chemtrac Systems Inc.
CH2M HILL
Clancy Environmental Consultants Inc.
Damon S. Williams Associates, LLC
Earth Tech Inc.
Ecolab Water Care Services
EMA Inc.
The Eshelman Company Inc.
Finkbeiner, Pettis, & Strout Inc.
Frontier Geosciences Inc.
ftn Associates Inc.
Gannett Fleming Inc.
GE Betz
Golder Associates Inc.
Greeley and Hansen LLC
The HACH Company
Hazen & Sawyer, P.C.
HDR Engineering Inc.
HNTB Corporation
HydroQual Inc.
Insituform Technologies Inc.
Institute for Environmental Technology & Industry, Korea
Jacobson Helgoth Consultants Inc.
Jason Consultants Inc.
Jordan, Jones, & Goulding Inc.
KCI Technologies Inc.
Kelly & Weaver, P.C.
Kennedy/Jenks Consultants
Komline Sanderson Engineering Corporation
Lawler, Matusky & Skelly Engineers, LLP
Limno-Tech Inc.
Lombardo Associates Inc.
Malcolm Pirnie Inc.
McKim & Creed
MWH
New England Organics
Odor & Corrosion Technology Consultants Inc. (OCTC)
ONDEO Degremont Inc.
PA Government Services Inc.
Parametrix Inc.
Parsons
Post, Buckley, Schuh & Jernigan
The RETEC Group
R.M. Towill Corporation
Ross & Associates Ltd.
Royce Technologies
SAIC Maritime Technical Group
Sear-Brown
Sierra Environmental Services
SYNAGRO
Tetra Tech Inc.
Trojan Technologies Inc.
URS Corporation
USFilter NATC
Vortechnics Inc.†
Wade-Trim Inc.
Weston Solutions Inc.
Woodard & Curran
WRc/D&B, LLC
WWETCO, LLC†

Industry

American Electric Power
BP
ChevronTexaco Energy Research & Technology Company
The Coca-Cola Company
Dow Chemical Company
DuPont Company
Eastman Chemical Company
Eastman Kodak Company
Merck & Company Inc.
ONDEO Services
Procter & Gamble Company
Severn Trent Services Inc.
Shell Global Solutions
Thames Water Plc
United Water Services LLC

† Charter Stormwater Subscribers

Note: List as of 1/03

WERF Product Order Form

As a benefit of joining the Water Environment Research Foundation, subscribers are entitled to receive one complimentary copy of all final reports and other products. Additional copies are available at cost (usually $10). To order your complimentary copy of a report, please write "free" in the unit price column. WERF keeps track of all orders. If the charge differs from what is shown here, we will call to confirm the total before processing.

Name ____________________ Title ____________________

Organization ____________________

Address ____________________

City __________ State __________ Zip Code __________ Country __________

Phone __________ Fax __________ Email __________

Stock #	Product	Quantity	Unit Price	Total
			Postage & Handling	
			VA Residents Add 4.5% Sales Tax	
			Canadian Residents Add 7% GST	
			TOTAL	

Method of Payment:

(All orders must be prepaid.)

❑ Check or Money Order Enclosed

❑ Visa ❑ Mastercard ❑ American Express

Account No. __________ Exp. Date __________

Signature __________

Shipping & Handling:

Amount of Order	United States	Canada & Mexico	All Others
Up to but not more than:	Add:	Add:	Add:
$20.00	$5.00*	$8.00	50% of amount
30.00	5.50	8.00	40% of amount
40.00	6.00	8.00	
50.00	6.50	14.00	
60.00	7.00	14.00	
80.00	8.00	14.00	
100.00	10.00	21.00	
150.00	12.50	28.00	
200.00	15.00	35.00	
More than $200.00	Add 20% of order	Add 20% of order	

*minimum amount for all orders

Note: Please make checks payable to the Water Environment Research Foundation.

To Order (Subscribers Only):

Log on to www.werf.org and click on the "Product Catalog."

Phone: (703) 684-2470
Fax: (703) 299-0742.

WERF
Attn: Subscriber Services
601 Wythe Street
Alexandria, VA 22314-1994

To Order (Non-Subscribers):

Non-subscribers may be able to order WERF publications either through WEF (www.wef.org) or IWAP (www.iwapublishing.com). Visit WERF's website at www.werf.org for details.

IWA Publishing's authorised EU representative for General Product Safety Regulations is Diane D'Arras, 15 rue Duret, 75116 Paris, France, e-mail: safety@iwap.co.uk.

Printed and bound by CPI Group (UK) Ltd, Croydon, CR0 4YY
12/05/2026
02108886-0005